# ANNALS OF SYSTEMS RESEARCH

W0193259

In the ANNALS OF SYSTEMS RESEARCH are published original papers in the field of general systems research, both of a mathematical and non-mathematical nature. Research reports on special subjects which are of importance for the general development of systems research activity as a whole are also acceptable for publication. Accepted languages are English, German and French.

Manuscripts should be typewritten and double spaced. Special symbols should be inserted by hand. The manuscript should not contain directions to the printer, these have to be supplied on a separate sheet. The author must keep a copy of the manuscript.

The title of the manuscript should be short and informative. An abstract and a mailing address of the author must complement the manuscript.

Illustrations must be added in a form ready for reproduction.

Authors receive 25 offprints free of charge. Additional copies may be ordered from the publisher's.

All manuscripts for publication and books for review should be sent to

A.F.G. Hanken
Technische Hogeschool Twente
Postbox 217
Enschede, The Netherlands

# ANNALS
# OF SYSTEMS RESEARCH

VOLUME 1, 1971

PUBLIKATIE VAN DE SYSTEEMGROEP NEDERLAND

PUBLICATION OF THE NETHERLANDS SOCIETY FOR SYSTEMS
RESEARCH

EDITOR:

B. VAN ROOTSELAAR

H. E. STENFERT KROESE N.V.  LEIDEN / THE NETHERLANDS

ISBN 978-90-207-0306-1     ISBN 978-1-4615-6446-1 (eBook)
DOI 10.1007/978-1-4615-6446-1

© H. E. Stenfert Kroese N.V./Leiden-The Netherlands

# PREFACE

This volume contains the texts of the lectures presented at the scientific meetings of 9 May 1970 and 23 January 1971 of the Systeemgroep Nederland. This society was founded on 9 May 1970 to promote interdisciplinary scientific activity on basis of a systems approach. It has its seat in Utrecht.

Officers for the year 1971:

President: A. F. G. Hanken, Technische Hogeschool Twente, Enschede.

Secretary: G. de Zeeuw, Department of Psychology, University of Amsterdam.

Treasurer: R. F. Geyer, Foundation for Interuniversitary Social Research, SISWO, Amsterdam.

All information about the society can be obtained from the Secretary.

Although this volume is entirely devoted to the lectures of the scientific meetings, future issues, to be collected into yearly volumes, are open to individual contributors. For details cf. p. II

The editor

# ADDRESSES OF AUTHORS

Buijs, B. G. F., Technische Hogeschool Twente, Postbox 217, Enschede.

Griend, P. C. van de, Nederlandse Economische Hogeschool, Burg. Oudlaan 50, Rotterdam.

Hanken, A. F. G., Technische Hogeschool Twente, Postbox 217, Enschede.

Leeuw, A. C. J. de, Technische Hogeschool, Insulindelaan 2, Eindhoven.

Maarschalk, C. G. D., Zomerluststraat 11, Haarlem.

Mantz, M. R., Trompenbergerstraat 7, Hilversum.

Meuwese, W., Technische Hogeschool, Insulindelaan 2, Eindhoven.

Rootselaar, B. van, Landbouwhogeschool, Afdeling Wiskunde, De Dreijen 8, Wageningen.

Wassink, E. C., Landbouwhogeschool, Afdeling Plantenphysiologie, Gen. Foulkesweg 72, Wageningen.

# CONTENTS

VII

# A SYSTEMS APPROACH TO PERSONAL
# AND GROUP FUNCTIONING

P. C. VAN DE GRIEND

It is the purpose of this article to show that systems thinking can be a useful tool in dealing with the problems that arise from the complex interrelations between person and group. A group of people consists of persons, but personal behavior is at least partly determined by group factors according to the classical Lewinian formula: $B = f(P, S)$ ($B$ = behavior, $P$ = person, $S$ = situation).

So the naive view in which the person is seen as the more fundamental entity and the concept of group is derived, a view which of course is consistent with the physical aspect of groups, does not hold true if we consider *psychological* functioning of persons and groups. It becomes apparent then that neither of the two can be seen as absolutely fundamental to the other. To give just one example: Bion, the English psychiatrist who wrote a classic on group phenomena recognizes three what he calls basic assumptions, modes of group emotionality, labelled: dependent, fight-flight and pairing. I shall return to these basic assumptions later on, but for the moment it is important to note that these states of group emotionality on the one hand have super-individual qualities, but on the other hand can only come into existence by individual contributions. Once they are there the individual is caught in, so to speak.

I shall not go into greater detail in advocating arguments for the view that person and group cannot be treated as wholly separate concepts, since this paper is not about the substantial or may be even philosophical question stated as the nature of the object if we are dealing with person-group phenomena, but about certain consequences. It seems clear that, if this view is taken, be it only in a hypothetical manner, what is necessary is a conceptual framework that can be applied to both personal functioning and group functioning. If such a framework can be developed a fruitful basis exists for an integrated science which comprises personality theory and group dynamics.

Of course attempts have been made to strive towards this kind of integration, beginning already with Freud, who applied his analytical theory to groups. The importance of social factors in personality functioning is even more stressed in Adlerian psychology. There are a number of more recent personality theorists in whose work social interaction plays an important or even a central role, such as Horney, Fromm, Moreno, Sullivan. On the other hand Lewin, who can be considered to be the founder of group dynamics, did not start from a completely social animal conception of man, a way of thinking about group phenomena, in which there is no room for personality factors. However in all these cases there is no complete integration at a conceptual level between personal behavior and group behavior. Of course it can not be the aim of this paper to develop a theory which covers all or even a great deal of phenomena in personal and group functioning. Much more modestly however it seems possible to work out at a rather abstract level some notions, that fall within the broad scope of systems thinking and which can be seen as finding support from other work in the fields of both personality theory and group dynamics.

### Autonomous and homonomous trends

A point of departure can be found in the work of Andras Angyal, whose *Foundations for a science of personality*, as early as 1941 goes into problems of systems thinking. That means Angyal does not evade the problems of logic and analysis, e.g. of relationship between elements within a system, which arise if one presents a very holistic approach. This holistic approach focuses on the concept of biosphere, derived from the German *Lebenskreis*. 'The biosphere includes both the individual and the environment, not as interacting parts, not as constituents which have independent existence, but as aspects of a single reality which can be separated only by abstraction' (p. 100). Within this biosphere Angyal recognizes two fundamental principles: the trend towards increased autonomy and the trend towards homonomy.

Increased autonomy is a descriptive term which indicates the way according to which the life process takes place. One could speak of self-expansion of a subsystem within a broader context, within a system. Angyal in his theory is not so much concerned with the relation between individual and group as I am in this paper, so applying his notion to group functioning one might say that the trend towards increased autonomy can be found in the individual in a group setting, but also in a sub-group with regard to a larger group. It is clear, that stated this way the trend towards increased autonomy is a rather abstract indication for a great number of phenomena which show more variations, the more one approaches concrete reality. Self-determination, agressiveness, the urge for mastery, expansion, drive for action, for acquisition, for exploration, also drive for integrity (resisting being dominated) may all be, according to Angyal, aspects of the general trend towards increased autonomy. In pointing out this tendency the individual organism, the sub-group, or sub-system is considered as if it were standing alone facing its environment but not being part of this environment, part of a broader system, be this social group, culture, religion or nature (even at the somatic level interindividual and superindividual processes take place).

The trend to be in harmony with the broader system is named by Angyal the trend toward homonomy. Again this trend manifests itself in a great number of ways. Individuals e.g. submerge themselves to superindividual wholes such as groups, already at a preverbal, emotional level. The goals of the homonomous trends are sharing, participation, union.

In concluding this paragraph it is important to note that the two tendencies do not operate separately, apart from each other. One seldom finds pure manifestations of either autonomous or homonomous trends, but the usual picture is a complex intermingling of both. One might even say, that the trends can only be distinguished at an abstract level, but in reality the one cannot exist without the other. There is no self-expansion possible if there is absolutely no participation in an environment, there is no surrender, no relationship without an entity which surrenders or relates. Consequently both trends are not necessarily antagonistic which at first sight they may seem, and however much in complex situations the trends may operate in contradictory ways. One should remember however Angyal's concept of biosphere as an integrated whole in which different principles operate.

So far I have followed Angyal rather closely. Although Angyal says (p. 179) that typological classifications might be derived from his theory, he does not do so himself. Exactly in this regard I want to point out some possibilities which will bring me back to the starting point of the article: a conceptual framework for personal as well as group functioning.

**Deficiency and compensation**

A basis for typological differentiation can be found by combining the idea of autonomous and homonomous trends with another important notion about personal and group functioning, which can be spotted in many different forms. I think e.g. of Maier's differentiation between frustrated and motivated behavior, of Roger's idea's about congruency and the optimally functioning person, and of research in connection with the concept of stress. In this brief article I shall leave out the differences between these several concepts but only pay attention to the similarity. It seems to me that two conditions can be recognized in which autonomous as well as homonomous trends operate: a condition in which these trends are blocked, frustrated, or said otherwise a condition in which they are functioning deficiently and a condition in which their functioning is optimal, not hampered in some way or another from within or without. I suggest to indicate these conditions by respectively a minus ($-$) and a plus ($+$) symbol.

Again it is clear that these symbols are rather abstract. But they do point to emotional factors; what follows are essentially symbolic descriptions of *affective structures*. They may refer to very different things in different situations; they indicate absolutely nothing about e.g. causes of frustration. Of course it is a very different thing whether a neurotic person is blocked in his tendencies towards autonomy or the emotional atmosphere in a certain group setting does not allow for the expression of autonomy by individual members. In the latter case the blocking of autonomous trends is a property of the total system which does not originate from the personalities of the individuals considered separately. Both conditions however may be indicated as $A-$. In a similar way a state of isolation, unrelatedness, non-participation of the subsystem in its environment may be referred to as $H-$. Now we have to take into account that in any subsystem-system relationship both autonomous and homonomous tendencies operate. As each of them can be in the state of optimally functioning or stress, blockage, we find four possible combinations, namely: $A-H+$, $A+H-$, $A-H-$ and $A+H+$.

In the next paragraph I shall deal more in the concrete with the way these typological differencies manifest themselves by relating them to certain clinical distinctions developed by other authors. First however I want to continue the argument about the nature of these symbolic descriptions. It is clear that they can only be global descriptions of certain ways of personal or group functioning. The main point is not that they are rich in meaning but that abstract classifications based on fundamental differences are possible. I am even more concerned with showing the possibility of a certain type of classification than with any claim that the type under consideration is absolutely correct. To put it even sharper: I have published the model which I am sketching here briefly before in a more elaborate form and at the present time I am working out an alternative form using a partly different way of conceptualization, which made me recognize five instead of four modes.

As a consequence of the foregoing the label $A+H-$ or any of the others does not reveal the manifold and intricate psychological mechanisms which operate under such headings. From the symbolic definitions the conclusion might be drawn that either autonomous or homononous trends can be deficient completely independent from the other ($A-H+$ and $A+H-$). In fact this independency does not exist: it has been said already that both tendencies influence each other which they do as well in a condition of stress and frustration. So if the expression of autonomous trends is blocked in some respect the expression of homonomous trends also suffers: the $+$ and $-$ symbols only locate more or less precisely the origin.

We can follow the processes involved a little closer by using the notions of repression and compensation. Tendencies that by some inner or outer cause are repressed try to find an outlet which gives rise to different kinds of compensating mechanisms. Compare with a river prevented to follow its main stream, where small branch-rivers develop. In a dependent condition, where autonomy is repressed, counter-dependency may occur; that means: the person or subgroup is critically resisting, opposing to the environment but in a non-asserting manner. No real autonomy is involved; it is only a compensating attempt, bound to fail to reach autonomy. Symbolically we van describe the reaction as: $A - H + \rightarrow a + h -$. In a similar fashion counter-balancing mechanisms can be supposed to take place in other conditions. Blocked homonomous trends are compensated with superficial participation in the broader system, which does not fundamentally alter the state of isolation. In symbolic terms: $A + H - \rightarrow a - h +$. It will be evident, that the integrated, non-frustrated condition $A + H +$ will show no compensating reactions: it does notsuffer from inner contradictions. For the $A - H -$ condition we can postulate: $A - H - \rightarrow a + h +$. That means: there is superficial autonomy as well as adaptation to the system. One can think e.g. of the benevolent leader, who at a more fundamental level does not relate at all to his followers. Neither is there real autonomy involved: the reader will remember the concept of autonomy as self-expansion *within* an environment. Now the main characteristic of a certain type of authoritarian leadership is exactly the opposite: autonomy is based not on *self*-expansion within an environment but on complete suppression of the total environment. Stated in different terms: the basis for autonomy is not situated in the subsystem, but in a deficient way in the system.

### Clinical evidence and applications

After having outlined the possibility of typological differentiations based on subsystem-system relationships the next question is if these theoretical reflections can be supported with clinical and empirical evidence. It is the author's conviction that there are studies of different kind, clinical, empirical research or more theoretical, from which findings can be reformulated or reinterpreted with the help of the considerations above. It is beyond the scope of this article to undertake this enterprise to the full extent. I shall restrict myself to two authors: Karen Horney and Bion and indicate briefly how there work can be related to what I have said. In particular I am concerned to show the power of the conceptual framework that I developed for uniting, for integrating points of view which were developed in different fields.

Karen Horney, founding herself on clinical experience, developed a theory of neurosis, in which she recognized three different types of inner conflict, each leading to deficient functioning. These neurotic structures, called respectively: moving toward, moving away and moving against all show inner contradictions.

The moving toward person does not reach autonomy; this type needs to be liked, loved, he cannot feel separate from symbiotic relationships. He tries automatically to live up to the expectations of others, showing complying, appeasing behavior. Agressive impulses are repressed but cannot completely be prevented from asserting themselves. They do so in ways that fit into the structure, more or less secretly, or if accumulated in a sudden outburst.

Quite on the contrary the moving away type, who acts under the appeal of freedom, has a compulsive need for detachment. His main concern is not to get emotionally involved with others. Anxiety is aroused if their environment intrudes on them. Autonomous tendencies are not blocked with this type: on the contrary, the core of the personality is to be understood from ideas of self-sufficiency, of 'splendid isolation'. Well developed is the capacity to act as an onlooker not only on others, but also on himself. The compensating behavior which I indicated is mentioned by Karen Horney as the possibility to get along with others superficially, just because one is not really interested.

Karen Horney's third tendency moving against people implies the rejection with violence of the softer human sentiments, like friendship, love, affection, sympathetic understanding. There is compulsive agressiveness that stems from the feeling that the world is an arena where: *homo homini lupus*. Though superficially this type shows fewer inhibitions than the other two, this might be due to (and not be the credit of!) our civilization. But as Karen Horney says in fact there is a choking of all feeling and great inner weakness. And it is especially this aspect which made me call this tendency $A-H-$. There can however be some discussion on the correctness of doing so, because on the other hand the strong urge for mastery, for excellence, for exploiting and outsmarting other does not in the first place make one think of repressed tendencies for autonomy. In a revised and more elaborate version of the theory under consideration I hope to deal with this problem more adequately.

Coming to the main argument of this paragraph now to compare Horney's three types of basic conflict with Bion's three basic assumptions. For two of the three the similarity is quite evident. The dependency phenomena that Bion describes in groups, where all good is expected from some authority figure are fundamentally a moving toward attitude. Agressiveness as well as weakness are manifested in Bion's fight-flight basic assumption, well to be compared with moving against. However at first sight Horney's isolated, detached structure and Bion's pairing group culture seem more or less opposite. If a group is passing from one of the other two cultures into a pairing assumption individuals are not avoiding but approaching each other. Thus the group is splitting up in pair relationships or at least in a number of smaller subgroups. But now if we do not consider the situation from the individual-group point of view but from the subsystem-system angle Horney's and Bion's condition do not differ so much. Pairing means that subsystems become detached from the total system. It might be so that avoidance of integration in the total system is the underlying force which leads to pairing and not in the first place 'moving towards' between individuals. They approach each other in order to isolate together from something threatening in the environment. The systems approach in this case makes plausible an interpretation that is different from Bion's. After having illustrated the possibility of applying the typological interpretations to persons as well as to groups it will be obvious that there is a wide potential for working with the system. Not only can one try to interpret relations between groups in these terms, one can also attempt to see cultural differences in the light of fundamental assumptions. Elsewhere I described certain cultural differences between the USA and the Netherlands in terms of $A+H- \rightarrow a-h+$ and $A-H+ \rightarrow a+h-$ respectively. I shall conclude this article with an application on leadership-groupclimate types.

The traditional division is tripartite; recognizing authoritarian, democratic and laissez-faire climates. The classification model, outlined briefly above, leads to more distinctions. There are 16 combinations possible if we start from four types of groupclimate and (the same) four types of leadership. It is easy to understand that not all 16 combinations are practically relevant (e.g. an $A+H+$ integrated group with an exploiting $A-H-$ leader). In the realm of democratic leadership alone however one comes to finer discriminations. I mention three forms which all can be called democratic.

| Group: | Leader | |
|---|---|---|
| $A+H- \rightarrow a-h+$ | $A+H- \rightarrow a-h+$ | individualistic democratic |
| $A-H+ \rightarrow a+h-$ | $A-H+ \rightarrow a+h-$ | social democratic |
| $A+H+$ | $A+H+$ | integrated |

In all three cases leader and group act on the same assumptions. The third, integrated form is more or less utopian: there is no frustration of autonomous and homonomous trends with leader and group. In the individualistic form the emphasis is on freedom, friendly interaction, but there is some repression of common interests, in so far as they conflict with individual freedom. Contrary in the social democratic atmosphere common interests, goals, etc. are stressed but then individual freedom is more under pressure, giving rise to compensating critical oppositon $(a+h-)$.

As a final remark I want to repeat that in this paper many complications have been left out, the aim being only to illustrate a certain way of thinking about group functioning.

## References

Angyal, A., *Foundations for a science of personality*. Cambridge (Mass.) 1958 4th ed.
Bion, W. R., *Experiences in groups and other papers*. London 1961.
Griend, P. C. van de, *Leren doceren*. Groningen 1970.
Horney, K., *Our inner conflicts*. London 1946 (4th ed. 1969).
Horney, K., *Neurosis and human growth*. London 1951.

# SYSTEMS ANALYSIS AND BUSINESS MODELS

A. F. G. HANKEN AND B. G. F. BUIJS

## 1. Introduction

As organizations are becoming more complex there is an increasing need to investigate already in the planning stage the results of a number of alternative policies. The construction of a model is thereby essential. It can be assumed that every employee has in mind some more or less vague notion which serves as a model of his company. These models are implicit, they are of little use to someone else. Explicit models fairly descriptive of company performance are now in frequent use as a tool for the policymaker. The idea of an integral business model, which includes many aspects of company activities, is certainly not new. In the Anglo-American literature one encounters the name 'corporate model' [8], [2], in the French literature 'modèle global de l'enterprise' [1].

Conceptually the most simple type is a block diagram [3]. The blocks refer to functional areas of the enterprise, the arrows connecting the blocks indicate that there is a relationship of some kind. These models serve as a convenient way to structure a given system. Qualitative models provide a verbal description of a system by enumerating the important attributes and their relationships relative to a given problem [7]. A quantitative model is best suited to describe the dynamic behaviour of complex systems with feedback phenomena and many interactions between components [4].

In this paper we are interested in the systems approach as a tool to analyse complex systems such as business organizations. A systematic approach, by which we mean the construction of a general framework, has above all educational advantages as it serves to improve the quality and consistency of a curriculum in business education. The term systems analysis, now often used in literature, means different things to different people. We subscribe to the opinion that systems analysis is not a unique method or theory in the sense of the theory of differential equations. It is a generic term which includes a number of methods. At the highest level of abstraction systems analysis is often synonymous with the construction of a model of great generality such as the state-space model [9]. The general equations of this meta-model can be used to structure more specific mathematical models such as differential or difference equations [9], [6], [5].

The concept of systems analysis also includes a method to structure complex real world systems. Here we are in the realm of the empirical models. It is typical of the systems approach that the terminology and framework of these abstract models are applied to conveniently solve problems, existing in the real world, by means of a model construction. Finally, it is used in the methodological sense whereby the methods of construction and application of models are the main topics of interest.

It should be mentioned that these four areas of systems analysis cannot be strictly separated. In this paper, as stated, the emphasis is on the construction of a general framework which can be readily applied to a complex system such as a business organization.

## 2. The systems approach

It is essential for any systems approach that a model of some kind is involved. A model is, of course, a simplification as reality can never be adequately represented by this construct. However, it is important to build a general model which allows the inclusion of many variables and at the same time gives some insight into the basic processes of the given system.

In this context our approach may be briefly summarized as follows:

1. An organization model is considered as a network of submodels.
2. The structure of each submodel is represented by a modelcell; this point will be further discussed below.
3. The modelconstruction starts with a submodel directly related to a problem. The network is completed by the addition of other submodels which have as outputs one or a number of inputs of the initial submodel. This process continues till all the inputs of the submodels have as a point of origin either another submodel or the environment of the given system.

Now the question arises whether there is a general way of structuring a model or submodel. To this end it will be useful to make a distinction between five different types of variables:

1. The inputvariables which are independent variables. They cannot be controlled by management i.e. by a person or a group of persons.
2. The decision or controlvariables are also independent variables. They can within certain limits be controlled by management to optimize company performance.
3. The outputvariables are dependent variables. It is stressed here that they are not necessarily related to the physical output of a system, e.g., the flow of

finished products leaving a factory, rather they are the variables directly or indirectly related to a given problem.

4. The intermediary variables appear in the modelequations both in independent and dependent positions without time delay.

5. Finally, the state variables have similar characteristics, i.e., they also appear as independent and dependent variables but there exists a time delay which is sometimes infinitesimally small (differential equations).

It should be mentioned that overlap between variabletypes is possible, e.g., an outputvariable may double up as a state variable.

In the next section an example will be presented to illustrate the given material.

### 3. Example of a modelcell

A salesmodel will be discussed here. The object of this model is to predict the sales of a company during a given period. It is a highly simplified model used for illustrative purposes only. The assumption of a regular growth pattern without business cycles or seasonal trends is made, it is furthermore assumed that there is only one product involved and that the company's marketshare never exceeds 80 %.

The modelequations are as follows:

$$MK(t) = a + bMK(t-1) \tag{1}$$

$$DM(t) = c - dPz(t)/Po(t) + e[V(t)/\{V(t)+Vo(t)\} - MA(t-1)] \tag{2}$$

$$MV(t) = MA(t-1) + DM(t) \tag{3}$$

$$MA(t) = MV(t) \text{ provided } MV(t) \leq .80 \tag{4a}$$

$$MA(t) = .80 \text{ provided } MV(t) > .80 \tag{4b}$$

$$O(t) = MA(t) \cdot MK(t). \tag{5}$$

The symbols have the following meaning:

$MK$ = total sales per period for the marketsector
$DM$ = increase per period of the company's marketshare
$Pz$ = salesprice of the company's product
$Po$ = average marketsector price
$V$ = salescosts per period related to the company
$Vo$ = total salescosts per period of all competitors combined
$MA$ = marketshare of the company
$MV$ = defined by the third equation
$O$ = companysales per period
$a, b, c, d$ and $e$ are constants.

The classification of variables outlined in the previous section gives rise to the modelstructure of figure 1.

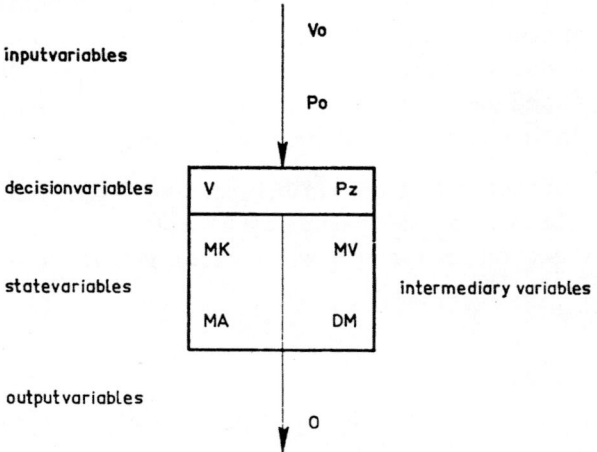

Fig. 1. Modelcell of a salesmodel.

It is immediately apparent that there are two input-, two decision-, two state-variables and one outputvariable. The quartet $(k, l, m, n)$ indicates the number of these variables, thus, in this case $(2, 2, 2, 1)$. It follows that there are 2 degrees of freedom (counted by the number of decision variables) and that the model is of second order because it has two state-variables. There are also two intermediary variables which are of little theoretical consequence as they can be eliminated, though this will complicate the equations to some extent. The number of equations equals the total number of state, intermediary- and output-variables, in this case five, whereby the equations for MA are counted as one. The modelcell representation has the advantage of the possibility of including many variables and that it relates to the state-space approach frequently used in systems analysis. It may be considered as a first step toward a quantitative model as it is not too difficult to write down a set of general modelequations if linearity and deterministic equations can be assumed. Conversely, the modelcell can also be used as a 'short-hand' notation of an already existing quantitative model.

## 4. The integration of modelcells

The construction of a structural model now proceeds as follows. First, a goal, problem or point of departure should be stated. From here it should be possible to define a number of variables, the so-called goalvariables. These are the variables one is interested in because of the nature of the problem. If the problem is waterpollution it will be necessary to make this concept operational by specifying the variables which are indicative of the degree of pollution. These goalvariables are by definition the outputvariables of the first modelcell. If this modelcell has undesignated inputs, i.e., inputs which do not originate in the environment, one may proceed to the next modelcell, where it is understood that these inputvariables are identical with the outputs or at least with some of the outputvariables of the second modelcell. This proces continues till the network is completed, which means that all 'open inputs' originate in the environment (figure 2).

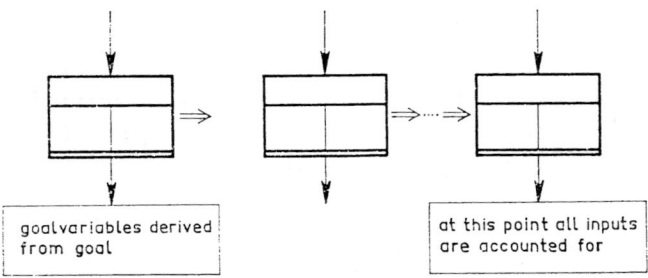

Fig. 2. The construction of a structural model.

It is important to tag the input- and outputvariables so that their points of origin and destination, i.e. one of the modelcells or the environment, are clearly indicated. By this method one avoids the drawing of cross-connections which would otherwise obscure the structural diagram.

## 5. Application of the systems approach

The above outlined method was used to plan a curriculum in business education. The basic question was where to start. It was decided by a group of eight persons, who worked intermittently on this project, that the balance sheet was the most promising point of departure as it includes many key variables of an enterprise. The balance sheet and the profit and loss account can be represented by a submodel, where most variables appear as inputs. They refer to various other submodels, e.g., the creditor account refers to a financial submodel. By working out a hypothetical case the structure outlined in figure 3 emerged. In this case, there are six submodels, representing six functional areas of enterprise, i.e., the sales-, planning-, investment/financing-, personnel-, production- and bookkeeping submodels. The most important links connecting the various submodels are listed below:

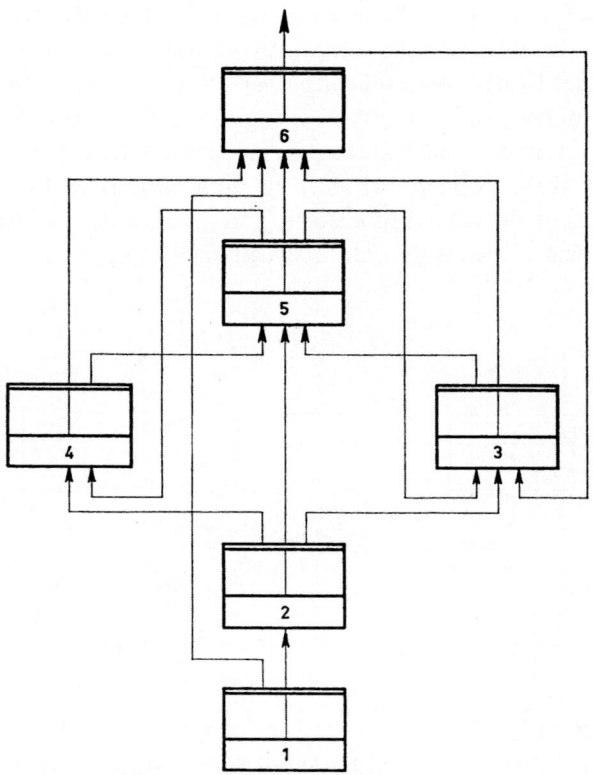

Fig. 3. Structure of an integral business model, consisting of six modelcells.

| | |
|---|---|
| 1. sales model | 4. personnel model |
| 2. planning model | 5. production model |
| 3. investment/financing model | 6. bookkeeping model |

| From | To | Connections between submodels |
|---|---|---|
| *i* | *j* | |
| 1 | 2 | sales per period |
| 1 | 6 | salescosts per period |
| 2 | 3 | long term salesforecast |
| 2 | 4 | long term salesforecast |
| 2 | 5 | short term salesforecast |
| 3 | 5 | production capacity |
| 3 | 6 | cost of capital |
| 4 | 5 | labour capacity |
| 4 | 6 | wages and salaries |
| 5 | 3 | capital demand |
| 5 | 4 | employee demand |
| 5 | 6 | production costs, distribution costs, sales income |
| 6 | 3 | retained profits. |

In the design phase submodel 6 was the first and submodel 1 the last one to be constructed. It is emphasized here that this structure is far from complete.

Submodels may be added or omitted, others may be split up into a number of submodels, depending on the type of company. Also the environmental influence has been omitted. It is also clear that this method of modeldesign should be applied anew for any given organization.

At present this approach is being used in the planning and sequencing of courses in a business education program. The main topics of interest appear as submodels in the structural diagram, the significance of the links between these submodels can also be traced. In this way it is possible to construct a fairly complete program with few overlapping areas.

Another advantage of this method is that the groundwork has been laid for a planning and decision procedure and a management information system. In this respect it is convenient to make a distinction between operational and structural decisions. The first type is, in essence, an optimization procedure whereby the decisionvariables are adjusted to obtain an output in accordance with a given goal. A structural or strategic decision corresponds to a choice between submodels within a given functional area. In this context it is useful to introduce the term module which is defined as the collection of all submodels related to the same functional area, e.g., a production module is the collection of all submodels related to the production process. The planning process is then a sequence of choices between structural and operational alternatives.

A beginning has been made to verify this approach by applying it to an existing medium sized company. A first indication is that the model is completely operational as all variables and parameters appearing in the integral model could be easily determined, measured or estimated. In our view, this is the result of the choice of the starting point, namely the financial statement of an enterprise. Work on this project is still continuing so that definite results are not yet available.

In conclusion it can be said that systems analysis is valuable as a unifying principle, as an instrument to construct general frameworks and as a methodological tool to approach real world problems. One should not, however, expect that an integral business model will completely replace the management functions. The problems existing in the real world are too varied and time dependent to rely completely on a model construct. In this respect man is certainly at an advantage, as he has the capacity to distinguish the essential from the nonessential in the problem solving process.

### Acknowledgement

The authors are indebted to a number of persons who have freely given their cooperation on this project. Especially Prof. ir J. M. L. Janssen, ir F. J. de Haan, drs F. Hendriks, dr P. G. Bosch, mr F. P. Fogaras en ir P. Melis are mentioned, as they have given much assistance and advice in the construction of the sub-models. Of course, the final responsibility for any sins of omission or commission rests entirely with the authors.

## References

1. Bogaert, A., Le modèle économique global de l'entreprise, *Annales de Sciences Economiques Appliquées* 26e ann., Dec. 1968.
2. Boulden, J. B. and Buffa, E. S., Corporate models: on-line, real-time systems, *Harvard Business Review* July-August 1970.
3. Dale, E., The division of basic company activities. In: Litterer, J. A. (ed.), *Organizations* vol. I, p. 76 (sec. ed.). New York (N.Y.) 1969.
4. Forrester, J. W., *Industrial Dynamics*. Cambridge (Mass.) 1961.
5. Freeman, H., *Discrete-Time Systems*, New York (N.Y.) 1965.
6. Gill, A., *Introduction to the theory of finite-state machines.* New York (N.Y.) 1962.
7. March, J. G. and Simon, H. A., *Organizations*, New York (N.Y.) 1961.
8. Roots, W. K., Corporate Synthesis, *Works Management* Nov. 1966.
9. Zadeh, L. A. and Desoer, C. A., *Linear System Theory.* New York (N.Y.) 1963.

# ON MEASUREMENT, METHODOLOGY AND SYSTEMS

A. C. J. DE LEEUW [1]

## Introduction

One may ask whether the increased interest in systems has to be considered as the need of scientists to present continuously novel and personal approaches [7] or as an effort to find solutions for important problems in society. In our opinion both statements contain a nucleus of truth. Anyhow it is found that as to the second statement many are of opinion that more and more complex problems appeal increasingly to different disciplines. The call for interdisciplinary studies becomes louder. It seems therefore plausible that Bouldings' opinion on the necessary communication between disciplines is completely applicable [2]. In cases this communication takes place in a particular way. We mean the fact which is to be appraised that scientists leave their original field of study to devote themselves entirely to a different discipline.

Probably the study of phenomena as a system is new exactly because the concept of a system is used as a methodological tool. We return to this in the section on methodology. Apparently the need of scientists to present novel approaches is not appraised by Koontz. According to him it has led to an enormous confusion. On the section on methodology we will argue against a negative judgement on the diversity of approaches.

Already from the title of [1]: general systems theory: a new approach to unity of science, appears that the striving for unity of science is not new. It is an age-old dream [1]. In fact the unity of science-movement [13], [14] tried in different ways to promote the integration of science. Von Bertalanffy objected against the so-called reductionism which roughly came down to the opinion that all phenomena are reducible to physical phenomena. This criticism of the unity of science-movement although essentially correct, does not cover all aspects because the encyclopaedia produced by the movement [13] shows in fact that reductionism is considered as only one of the efforts to promote the unity of science.

[1] The author wishes to thank professor B. van Rootselaar for valuable comments on an earlier draft of this paper.

One might assert that General Systems Approach just like the Unity of Science movement propagates a unity of method and a unity of language. The difference resides in the starting-point of considering phenomena as systems and in the opinions about reductionism.

We hope to have suggested in the preceding that General Systems Approach has some important features which partly find their origin in the Unity of Science movement, to wit:

1. The unity of scientific method.
2. The unity of origin.
3. The unity of language.

We shall try to point out in the following a certain similarity between measuring and the construction of models of empirical phenomena. We start with the methodological starting-points to continue with the methodology of measurement and end with an exposition of our conception of systems theory.

A more general remark may be in order. Intrusion of a scientist into a field of knowledge not his own carries an amount of uncertainty. Therefore attempts to build bridges between the different disciplines will only be successful if these efforts are accompanied by an attitude of mutual understanding of the technical difficulties involved.

### Methodological considerations

Summing up the discussions in the literature of philosophy of science one may distinguish two main directions:

1. The opinion that all statements of an empirical science are based on observable and immutable facts.
2. The opinion that no reality exists outside our mental activity.

Expressed a little differently in terms of the relation between theory and empirics the two distinguishable opinions are: the theory is determined entirely by the facts and the facts are determined by the theory. We follow Popper [16], Feyerabend [3], and Kuhn [8] and take our position in between. It is briefly described by the following.

A theory consists of a sequence of sentences (cf. [5], [12]) constructed from a vocabulary consisting partly of a subset of the vocabulary of logic. The remaining terms are either primitive or derived ones. This set of theoretical terms is connected with empirical phenomena by operational definitions (rules of correspondence). It is not the case that an arbitrary theory and an arbitrary set of empirical facts can be brought in correspondence in a meaningful way, because in our opinion the theoretical terms determine to a certain extent the observed facts and conversely. So every new approach may reveal new facts and eliminate or change old facts and may therefore be more appropriate than the former approaches. For this very reason we do not agree with Koontz's opinion as we already stated in the introduction.

A systems approach may be considered as a theoretical startingpoint. It is a conscious choice to look at phenomena as systems. This has consequences for the facts to be observed. This is one of the reasons for us that attention has to be paid to the concepts of systems theory.

According to some authors (cf. [6]) a system is a well-defined mathematical object and systems theory largely a part of mathematics. It will be clear that we do not share this opinion. This implies that theories about mathematical structure are not of pre-eminent importance to us. However, the connection between these mathematical structures and the 'world' belongs essentially to systems theory. This implies the impossibility to restrict systems theory to mathematics. To indicate briefly the problems that arise in this connection, let $\omega$ be an empirical object, e.g. the present author. Properties or attributes may be assigned to $\omega$, such as weight, age, motivation, etc. We postulate that it is impossible to characterize $\omega$ completely by means of a finite number of attributes. On the other hand we assume the set of attributes under consideration to be finite.

From these assumptions follows that we cannot reasonably expect a one-to-one correspondence to exist between the empirical objects and the abstract ones.

Abstraction, the selective study of aspects of phenomena is characteristic for any empirical science.

An important method to connect theory with empirical phenomena is measurement, which we consider in the next section.

## Measurement

In discussions with workers in the behavioural sciences one not seldom encounters statements like: Physics has developed so quickly because there one has started with observations and measuring. Therefore behavioural sciences should concentrate more on methods of measurement.

However, in making such a statement one overlooks the fact that although in physics measuring is of great importance it is almost always induced by some theoretical conception. The interested reader may be referred to Kuhn [8].

Measurement is usually described as the assignment of numbers to phenomena according to some well-defined procedure. It is an important facet of the scientific method in all disciplines. And authors like Suppes and Zinnes [17] and Pfanzagl [15] have great merit for having established a theory of measurement applicable in different disciplines. We shall present the theory of Suppes and Zinnes here in outline, just to point out its main ideas.

If for the moment we start with describing measuring as the assignment of numbers to phenomena according to some well-defined procedure, there arise two questions. First one may ask whether and how far such an assignment is justified, which leads to the socalled representation problem. Second, one may ask in how far the assignment of numbers is unique, which leads to the uniqueness problem.

In order to formulate these problems more precisely we need a few logico-mathematical notions.

Definition: A relational system $U = \langle A, R_1, \ldots, R_n \rangle$ of type $(m_1, \ldots, m_n)$ is a set $A$ together with $n$ subsets $R_i \subset A^{m_i}$ ($m_i$-ary relations), where the $m_i$ are positive integers.

$A$ is called the domain dom $U$ of $U$, while $\langle R_1, \ldots, R_n \rangle$ is called its structure. Two relational systems are called similar if they have the same type.

### Examples

1. $U = \langle A, R \rangle$, where $A$ is the set of all humans, $A^2$ the set of all ordered pairs $(a, b)$, and $R \subset A^2$ is the binary relation defined by $(a, b) \in R$ if and only if $a$ and $b$ are either both male or female. Note that $R$ is reflexive, i.e. $(a, a) \in R$ for all $a \in A$, it is symmetric, because $(a, b) \in R$ implies $(b, a) \in R$ for all pairs $(a, b) \in A^2$, and $R$ is transitive, i.e. $(a, b) \in R$ and $(b, c) \in R$ imply $(a, c) \in R$ for all $a, b, c$ in $A$. Any relation that has these three properties is called an equivalence relation.

2. $U = \langle A, S \rangle$, where $A$ is as in the preceding example, while $S \subset A^2$ is defined by $(a, b) \in S$ if and only if $a$ is at least as old as $b$.

A system which is similar to the above is e.g. $\langle Re, \geqq \rangle$, where $Re$ is the set of real numbers and $\geqq$ the natural ordering $a \geqq b$.

3. A relational system involving a ternary relation is e.g. $V = \langle Re, R \rangle$, where $Re$ again is the set of real numbers, while $R \subset Re^3$ is defined by

$$R = \{(a, b, c); a, b, c \in Re \ \& \ a+b = c\}.$$

The following notion is crucial in the formulation of the representation problem.

Definition: A mapping $f: A \to B$ is called a homomorphism from the relational system $U = \langle A, R_1, \ldots, R_n \rangle$ to the similar system $V = \langle B, S_1, \ldots, S_n \rangle$ if $(a_1, \ldots, a_{m_i}) \in R_i$ implies $(f(a_1), \ldots, f(a_{m_i})) \in S_i$ for all $a_1, \ldots, a_{m_i} \in A$ and $i = 1, \ldots, n$.

If the homomorphism $f$ is one-to-one it is called an isomorphism.

In order to connect the collection of phenomena with the numbers which are to be assigned we distinguish two kinds of relational systems. On the one hand the empirical relational systems, which are relational systems whose domains are collections of identifiable objects in the environmental world. On the other hand numerical relational systems which have for their domains subsets of the set of real numbers.

Now essentially the process of measuring is the construction of a numerical relational system $N$ to some given empirical relational system $U$, such that $N$ is a homomorphic image of $U$. The representationproblem actually is the problem of showing some proposed $N$ to be homomorphic (or isomorphic) image of $U$.

In order to formulate the uniquenessproblem we need a few more notions. First a numerical relational system will be called a full system if its domain is the entire set of real numbers $Re$. Next the relational system $V$ will be called a subsystem of $U$ if the domain of $V$ is a subset of that of $U$ and its relations are the restrictions of those of $U$ to the domain of $V$.

Now we arrive at the essential notion of a scale.

Definition: The triple $\langle U, N, f \rangle$ is a scale if $U$ is an empirical relational system and $f$ is a homomorphism of $U$ onto a subsystem of the full numerical relational system $N$.

The uniquenessproblem is the question in how far $f$ is uniquely determined by the choice of $U$ and $V$. This question leads to the distinction of different types of scales, distinguished by the freedom there remains for the choice of $f$.

An example illustrating the general situation is found in the measurement of temperature. The structure of the empirical system arises from comparison of objects. To measure temperature different functions $f$ may be chosen, the three scales introduced by Celsius, Réaumur and Fahrenheit result from three different choices of the function $f$, which however are interrelated, and can be converted into each other.

Uniqueness of $f$ in the possible scales for $U$ is of importance for the meaning of the numbers which $f$ assigns to objects of $U$. If e.g. we find $f(a) = 3f(b)$ and using a different function $g$ we find $g(a) \neq 3g(b)$, it is evident that we cannot conclude anything about the connection between the properties of objects $a$ and $b$ we are measuring.

In the following we shall see that the restrictions involved in measurement can be brought out clearly in terms of the notion of admissible transformations.

Let $Hom(U, N)$ be the set of homomorphisms of $U$ into $N$ (i.e. onto a subsystem of the full sytem $N$) then there is a one-to-one correspondence between $Hom(U, N)$ and the set of scales $S(U, N)$, for if $f \in Hom(U, N)$, then $\langle U, N, f \rangle$ is a scale, and conversely by the definition of scale.

Now any real function $\phi$, which composed with some $f \in Hom(U, N)$ again yields an element of $Hom(U, N)$ is called an admissible transformation, because it induces the conversion of one scale into another.

For if $\phi$ is an admissible transformation and $g = \phi \circ f$, i.e. $g(a) = \phi(f(a))$

for all $a \in U$, and $f \in Hom(U, N)$ then $g \in Hom(U, N)$ i.e. scale $\langle U, N, f \rangle$ goes over into scale $\langle U, N, g \rangle$.

The set of admissible transformations corresponding to $U$ and $N$ is denoted by $A(U, N)$.

If the set $A(U, N)$ consists of one element, this has to be the identity transformation $e(x) = x$ for all $x \in Re$. Then $Hom(U, N)$ has at most one element, and if $f \in Hom(U, N)$ then the corresponding scale $\langle U, N, f \rangle$ which in this case is unique, is called an *absolute scale*. Scales may be classified by their sets of admissible transformations. Examples of scales which are actually used are the *nominal scales*, for which $A(U, N)$ is the set of one-to-one transformations, *ordinal scales*, for which $A(U, N)$ is the set of isotone transformations (i.e. transformations $\phi$ such that $(y-x)(\phi(y)-\phi(x)) > 0$ for all $x \neq y$), *interval scales*, for which $A(U, N)$ is the set of transformations of type $\phi(x) = \alpha x + \beta$, where $\alpha, \beta \in Re$ and $\alpha > 0$, and *ratio scales*, which are interval scales for which $\beta = 0$.

The notion of admissible transformations permits an elegant criterion for the meaningfulness of arithmetical expression referring to one particular scale ([17]). In fact Suppes and Zinnes define an arithmetical expression to be meaningful if its truthvalue is invariant under admissible transformations of scale. This may be illustrated by the following example:

1. Consider the expression

$$\sum_{i=1}^{n} x_i > \sum_{i=1}^{m} y_i, \text{ then for } \alpha > 0 \text{ we have}$$

$$\sum_{i=1}^{n} x_i > \sum_{i=1}^{m} y_i \leftrightarrow \sum_{i=1}^{n} \alpha x_i > \sum_{i=1}^{m} \alpha y_i$$

So if the $x_i$ and $y_i$ refer to a ratio scale for which the admissible transformations have the form $\phi(x) = \alpha x$, it follows that

$$\sum_{i=1}^{n} x_i > \sum_{i=1}^{m} y_i \leftrightarrow \sum_{i=1}^{n} \phi(x_i) > \sum_{i=1}^{m} \phi(y_i)$$

hence the expression is meaningful.

If however the $x_i$ and $y_i$ refer to an interval scale it follows from

$$\sum_{i=1}^{n} x_i > \sum_{i=1}^{m} y_i \leftrightarrow \sum_{i=1}^{n} (\alpha x_i + \beta) > \sum_{i=1}^{m} (\alpha y_i + \beta) + (n-m)\beta$$

that for $n \neq m$ there is a choice for $\beta$ such that on the one hand

$$\sum_{i=1}^{n} x_i > \sum_{i=1}^{m} y_i$$

may be true, while on the other hand

$$\sum_{i=1}^{n} (\alpha x_i + \beta) > \sum_{i=1}^{m} (\alpha y_i + \beta)$$

i.e. the truth value of the expression is not invariant under admissible transformations, hence the expression is not meaningful, i.e. does not express a property of the objects (in $U$) under consideration.

If $m = n$ however the expression is meaningful.

In physics many quantities are measured on ratio scales. The class of meaningful expressions is more extensive than is the case e.g. with ordinal scales, because the set of admissible transformations in the former case is more restrictive than in the latter. This fact is a reason for looking for the strongest possible scales, i.e. for those with smallest possible set of admissible transformation.

## Systems

Although many definitions of system can be found in the literature, we feel the need for a new one, notwithstanding Koontz's negative judgement of such an attitude. The present definition is inspired by Hall and Fagen [4].

In our rather abstract exposition we use $\omega$ as object variable and notice that objects can be anything from symbols, to individuals, machines, departments, etc. A set of objects will be assumed to be finite and can be defined by enumeration, e.g. $W = \{\omega_1, \ldots, \omega_n\}$, or by specifying properties of its members $W = \{\omega : P_1(\omega) \wedge \ldots \wedge P_n(\omega)\}$.

For recognizable (measurable) attributes of objects we use small latin letters. The set of attributes of an object $\omega$ will be denoted by $X_\omega$, and it is assumed to be finite, so we may write $X_\omega = \{x_1, \ldots, x_n\}$. If for example $\omega$ is an individual and $a_\omega, m_\omega, s_\omega$ stand for his attributes age, motivation and salary, then $X_\omega = \{a_\omega, m_\omega, s_\omega\}$.

If $W$ is a set of objects, we define the set of attributes of $W$ as the union of the sets of attributes of its members, i.e.

$$X_W = \cup \{X_\omega : \omega \in W\}.$$

Attributes can take certain values, i.e. the attribute colour has red as a value, etc. We do not enter into the problem of assigning values to attributes which is a problem of measurement.

What we wish to formalize is that one set of attributes is related to a second one. To do this we need a few definitions.

Let $D(x)$ stand for the set of values of attribute $x$. Then if $X = \{x_1, \ldots, x_n\}$ we denote the direct product of $D(x_1), \ldots, D(x_n)$ by $\Gamma(X)$, hence

$$\Gamma(X) = D(x_1) \times \ldots \times D(x_n) = \times \{D(x); x \in X\}.$$

So the elements of $\Gamma(X)$ are $n$-tuples of values of the attributes $x_1, \ldots, x_n$ of $X$ in this order.

Now if $T \subset Re$ then the set of non-constant functions from $T$ into $\Gamma(X)$ is denoted by $D(T, \Gamma(X))$, so $g \in D(T, \Gamma(X))$ if $g : T \to \Gamma(X)$ and there are $t_1$ and $t_2$ in $T$ such that

$$g(t_1) \neq g(t_2).$$

If we interpret $T$ as time then $D(T, \Gamma(X))$ is the set of functions which assign at any moment $t \in T$ values to all attributes of $X$, while these values are not all the same throughout $T$, that means at least one of the attributes changes with time.

Now we are able to tell when two sets of attributes, $X$ and $Y$ are related.

MAIN DEFINITION: *The attribute set $Y$ is said to be related to the attribute set $X$, in formula $R\{X \to Y\}$ if to any $g \in D(T, \Gamma(X))$ there corresponds an $h \in D(T, \Gamma(Y))$, in other words $Y$ is related to $X$ if there exists a mapping $\Phi: D(T, \Gamma(X)) \to D(T, \Gamma(Y))$.*

We illustrate this definition by an example. Let $\omega_1$ and $\omega_2$ be individuals, and we consider their salary $s$ and their motivation $m$ for work. Then $X_1 = X_{\omega_1} = \{s_{\omega_1}, m_{\omega_1}\}$ and $X_2 = X_{\omega_2} = \{s_{\omega_2}, m_{\omega_2}\}$. Let the possible values of $s$ be 50, 75, 100, 150, and suppose these values are open to both $\omega_1$ and $\omega_2$, i.e. $D(s_{\omega_1}) = D(s_{\omega_2}) = \{50, 75, 100, 150\}$, further suppose $D(m_{\omega_1}) = D(m_{\omega_2}) = \{\text{low, high}\}$.

Now we study the object set $W = \{\omega_1, \omega_2\}$ during the time-set consisting of four moments $t_1 < t_2 < t_3 < t_4$, so $T = \{t_1, t_2, t_3, t_4\}$.

The behaviour of $\omega_1$ and $\omega_2$ during $T$ is given by functions $g$ and $h$, such that $g : T \to D(s_{\omega_1}) \times D(m_{\omega_1})$ and $h : T \to D(s_{\omega_2}) \times D(m_{\omega_2})$. So possible values for $g$ are e.g. $g(t_1) = (50, \text{low})$ and $g(t_4) = (150, \text{low})$.

Similarly for $h$.

The entire behaviour of $\omega$, with respect to $s$ and $m$ is given by the four points in the salary-motivation-plane (cf. fig. 1 on p. 26),

e.g. $g(t_1) = (50, \text{low})$
$\quad g(t_2) = (75, \text{high})$
$\quad g(t_3) = (100, \text{high})$
$\quad g(t_4) = (150, \text{low}).$

If to such functions $g$ there correspond functions $h$ for $\omega_2$ which likewise show changes through time, we have reason to assume that the behaviour of $\omega_2$ is not quite independent of that of $\omega_1$ and therefore we say that $X_{\omega_2}$ is related to $X_{\omega_1}$.

Evidently our main definition induces the following additional definitions. Synonymous with $R\{X_1 \to X_2\}$ is $R\{X_2 \leftarrow X_1\}$. Further the sets $X_1$ and $X_2$

are interrelated, if one is related to the other and conversely, i.e.

$$R\{X_1 \leftrightarrow X_2\} = R\{X_1 \rightarrow X_2\} \wedge R\{X_2 \rightarrow X_1\}.$$

It is also useful to have the formula $R\{X_1 \nleftrightarrow X_2\}$ expressing the fact that $X_2$ is related to $X_1$ but not conversely. It is defined by

$$R\{X_1 \nleftrightarrow X_2\} = R\{X_1 \rightarrow X_2\} \wedge \neg R\{X_2 \rightarrow X_1\},$$

where $\neg R$ is the negation of $R$.

The following connection between two sets of attributes seems to be the most useful one:

$$R\{X_1 ; X_2\} = R\{X_1 \rightarrow X_2\} \vee R\{X_2 \rightarrow X_1\}.$$

If the sets $X_1$ and $X_2$ are the sets of attributes belonging to (single objects or) sets of objects it is useful to speak of this relation in terms of these sets of objects or the objects themselves, therefore we say for sets $W_1$ and $W_2$ of objects that they are related if their sets of attributes are related in the above sense.

So we introduce as abbreviations

$$R\{W_1 \rightarrow W_2\} = R\{X_{W_1} \rightarrow X_{W_2}\} \text{ and } R\{W_1 ; W_2\} = R\{X_{W_1} ; X_{W_2}\}.$$

In particular for individual objects $R\{\omega_1 \rightarrow \omega_2\} = R\{X_{\omega_1} \rightarrow X_{\omega_2}\}$, etc.

The thus introduced term $R\{A; B\}$, where $A$ and $B$ are sets of objects enables us to define the notion of a system.

SYSTEMS DEFINITION: *A set $W$ of objects is called a system if for all subsets $A$ of $W$ different from $\emptyset$ and $W$ we have $R\{A; W/A\}$.*

Note that $\emptyset$ is used for the void set, and $W/A$ means the subset of elements of $W$ that do not belong to $A$, i.e. $W$ intersected with the complement of $A$.

A system therefore is a set $W$ of objects with a set of relations and attributes defined for these objects such that any proper part of $W$ is related to the rest of $W$ or conversely the rest related to the proper part in the sense of our main definition.

Usually non-formal definitions of systems take into account some environment which is supposed to interact with it or at least is relevant to it. As a formal definition of environment $E(W)$ of a system $W$ we propose

$$E(W) = \{\omega; \omega \notin W \text{ \& } R\{\omega; W\}\}$$

and we distinguish closed systems, being systems $W$ for which $E(W) = \emptyset$ (void set) and open systems, being systems for which $E(W) \neq \emptyset$.

Once we have our definition of system we may continue to consider models of systems as homomorphic images and develop a systems theory. For attempts towards such a theory we refer to a number of reports which may be requested from the author.

In any case we hope to have given some indication of a possible connection of the theories of measurement with the construction of models of systems.

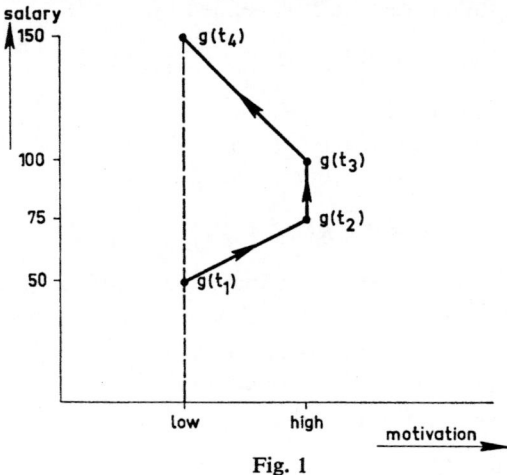

Fig. 1

## References

1. Bertalanffy, L. von, General system theory: A new approach to unity of science. *Human Biology* 23 (1951).
2. Boulding, K. E., General system theory: The skeleton of science. *Management Sc.* 2, 3 (1956).
3. Feyerabend, P. K., Explanation, reduction and empiricism. In: Feigl, H. a.o. (eds.), *Minnesota studies in the philosophy of science*, vol. III. Minneapolis (Minn.) 1962.
4. Hall, A. D. and Fagen, R. E., Definition of System. *General Systems*. I (1956).
5. Hempel, C. G., *Aspects of scientific explanation and other essays in the philosophy of science* New York (N.Y.) 1965.
6. Kalman, R. E. a.o., *Topics in mathematical systems theory*. New York (N.Y.) 1969.
7. Koontz, H., The management theory jungle. *J.A.M.* (Dec. 1961).
8. Kuhn, T. S., *The structure of scientific revolutions*. Chicago (Ill.) 1962.
9. Leeuw, A. C. J. de, *De systeembenadering van organisaties*, rapport no. 3, groep organisatie-leer, afd. der bedrijfskunde i.o. Eindhoven oktober 1968.
10. Leeuw, A. C. J. de, *De bestudering van systemen*, rapport no. 4, groep organisatieleer, afd. der bedrijfskunde i.o. Eindhoven november 1968.
11. Leeuw, A. C. J. de, *Enige aantekeningen over het begrip typologie en de implicaties daarvan op het ontwerpen van een typologie van produktiesystemen*, rapport no. 2., groep organistie-leer, afd. der bedrijfskunde i.o. Eindhoven oktober 1968.
12. Nagel, E., *The structure of science*. London 1961.
13. Neurath, O. a.o. (eds.), *International encyclopedia of unified science*, vol. I. Chicago (Ill.) 1955.
14. Oppenheim, P. and Putnam, H., Unity of Science as a working hypothesis. In: Feigl, H. a.o. (eds.), *Minnesota studies in the philosophy of science*, vol. II. Minneapolis (Minn.) 1958.
15. Pfanzagl, J., *Theory of measurement*. Würzburg 1968.
16. Popper, K. R., *The logic scientific discovery*. New York (N.Y.) 1960.
17. Suppes, P. and Zinnes, J. L., Basic measurement theory. In: Luce, R. D. a.o. (eds.), *Handbook of mathematical psychology*, vol. I. New York (N.Y.) 1963.

# THE USE OF ASPECT SYSTEMS IN A GENERAL MODEL FOR ORGANIZATIONAL STRUCTURE AND ORGANIZATION CONTROL

C. G. D. MAARSCHALK

## 1. Introduction

A better understanding of structure, operation and management of a business or any organization can be obtained by analysing it as a system, or rather a set of systems.[1]

This approach deals with the control of a business or other organization. In this age of automation it is essential to analyse how automation principles can be applied to certain parts of the management function, in fact to the control function; automation of only the processing of data is not enough.

Such analysis should lead to an objectively based method as to how to determine the control-information of the organization.

## 2. Terms and classification

There are many definitions of the concept 'system'. For our purpose, it is suitable to define a system as a set of objects (or elements), having internal relationsships, which functions (in an environment) as a unity or is being considered as such.

Everything which does not belong to a system is part of the environment. Usually the system has external relationships with the environment. It makes sense to limit the environment to be considered to that part which is or might be relevant to the system considered.

We can distinguish various kinds of relations, both internally and externally. Our view of a system pertains to physical entities and to phenomena which can be considered as a unity. The system then is created by the way of looking at something, it is a creation of the mind.

An object of a system can be a system in itself. Sometimes terms as subsystem or partial system then are being used; it means that the system has been cut

---

[1] More elaborately in Maarschalk, C. G. D., *Bedrijfsinformatica* (Business Informatics), in preparation.

*Annals of Systems Research 1* (1971), 27–41

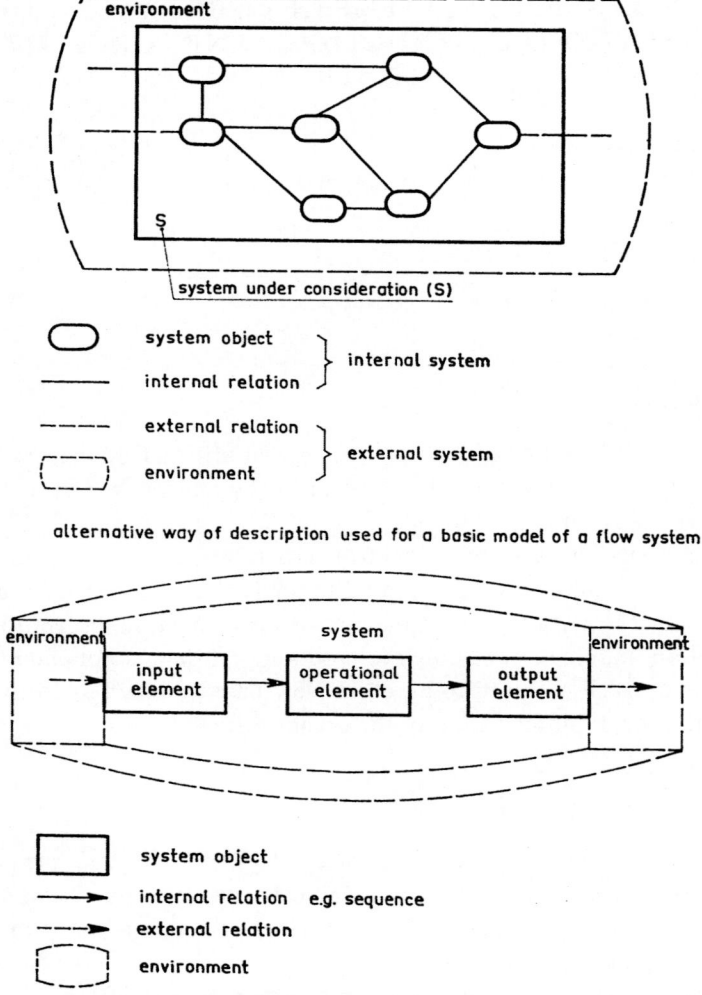

Fig. 1. General model of a system.

into a number of parts; they add up to the total system under consideration. In case of a business such parts often are compared with the departments or with the functions (manufacturing, sales, finance, etc).

Besides total system, also terms as integrated system or complex system are in use.

We are introducing the term *aspect system*, meaning a certain view at the total system, chosen by the contemplator. Aspect systems do not have to reflect physical or organizational order, they can be compared with 'layers' or 'strata' of a system.

All systems under consideration have one or more objectives or targets. That requires systems control in order to attain or at least aim for objectives or targets.

The systems involved can be controlled through feedback; it means that the output has an impact in some way on the input.

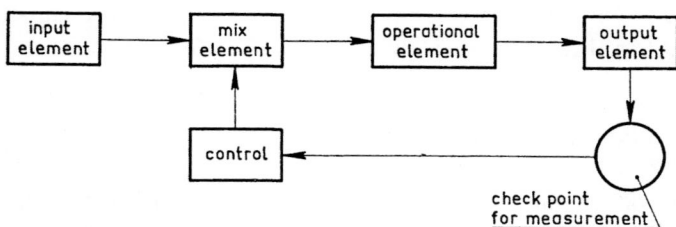

Fig. 2. System controlled by feedback

In the analysis of a system we distinguish between the concept of systems structure (the cooperation of objects and relations) and the concept of systems behaviour (the output, caused by structure and the input).
As to the behaviour we classify into two groups:

a. Systems, showing a characteristic connection between inputs and outputs, which can be designated as a flow. Characteristic relation includes a causal relevancy between inputs and outputs as to identities and quantities. It could be said that the input-output relation makes sense. Such systems are indicated as flow systems.
b. Systems, not showing such characteristic input-output relation. Then the system structure mainly determines the output. Such a system can condition a situation or some other system. We shall indicate such phenomena as conditioning systems or non-flow systems or as 'structural behaviour'.

As to the rate of predictability of systembehaviour systems can be grouped into classes. The classification used by Bosman[2] fits our purpose:

a. The real system, which is reality.
b, The symbolic system: a transition which results in sufficient predictability of behaviour for the purpose involved.
c. The iconic system: a transition not resulting in sufficient predictability of behaviour for the purpose involved.

Also a model is a transition of reality. Symbolic and iconic systems both are models. We need symbolic systems for a valid description of a business firm or other organization.

[2] Bosman, A., *Systemen, planning, netwerken.* p. 11 ff. Leiden 1969.

## 3. Presently used Business Models

There are a number of frequently used business models, which can be considered as systems, although they usually are not termed as such. Together they compose a picture of a business firm, which often is the basis for determining requirements for business information[3].

We want to investigate the validity and efficiency of the present situation.

The traditional set of models includes the following:

A. Models of physical business processes, serving the existing objectives of the organization. Examples: manufacturing processes and sales (marketing) processes.

B. Models of the financial/economic flowsystem, comprehending financial situations (reflected in the balance sheet), calculations of financial results and investments, etc.

C. Models of organizational structure, composed of a hierarchie of functions with complementary function- and task- descriptions.

D. Models of procedures, including collecting, processing, and filling of data with complementary formhandling.

E. Planning models of various kinds for manufacturing, finance, budgeting, etc.

We shall enter into a short analysis.

*Ad* A

Models of physical business processes. We illustrate our point by a simple manufacturing model[4],

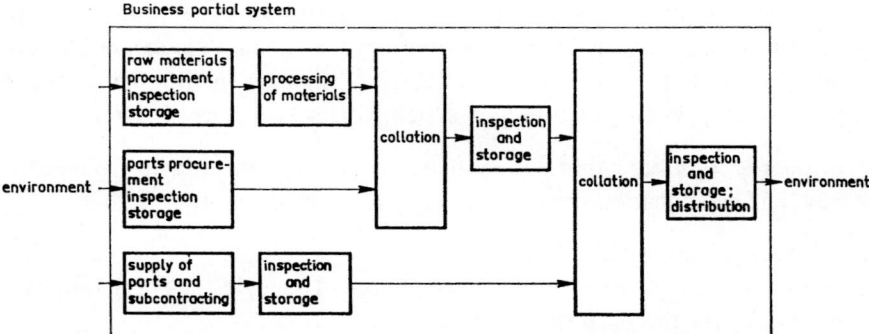

Fig. 3. Escampte of basic model of a manufacturing system. 3 System objects: physical parts and operations. 3 System relations: sequence, time and quantitative relations.

---

[3] Information: interpreted data, which make sense in a field of application. Information has a pragmatic character, supplying criteria for data selection.

[4] Manufacturing is a preferred term here. The word production is used by the economic sciences as contradiction to consumption and then includes e.g. storage, and sales distribution processes.

The manufacturing process is clearly a flowsystem: outputs are being determined in full or mainly by inputs. Inputs are raw material, labor and machine services, etc, outputs are final products and residuary, etc. It is a transformation system; as objects we can consider: processing activities and places but also the product being processed.

Description is usually made with aid of network- or flowcharttechniques. The principle structure is given in fig. 3. In this illustration only sequence is indicated (by arrows) as systems relation; in addition other relations, such as quantities and time, of course are possible.

Such Models often are used in problems of planning, time studies, capacity calculations, layout and routing, etc., mainly in organizational problems with a technical or economic-technical character.

*Ad* B

The financial model of a firm is formalised in the bookkeeping (accounting) system. The essential part of this system is the measurement of the transformation of investments in gross returns and costs.

This system processes situations, expressed in terms or money (balance sheets) and actual changes of the situation (net returns), with their qualitative and quantitative relations. Besides, this system can include more data and serve additional purposes, such as financial planning, cost analysis and auditing purposes. A set of accounts and a set of procedures serve as description techniques. The essence of this flowsystem is illustrated in fig. 4.

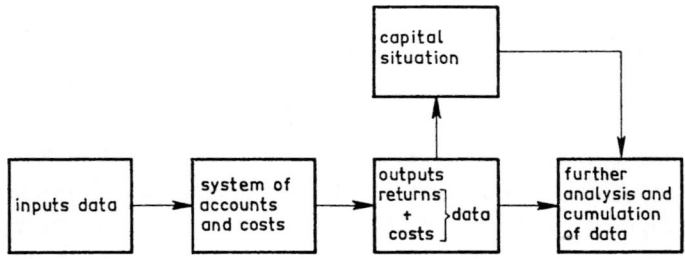

Fig. 4. Basic model of the accounting system

*Ad* C

The traditional description of a model of organizational structure is given in fig. 5, using a 'rake-pattern'. It includes a division of functions, based on hierarchal relation (vertical line) and specialisation (horizontal line). Working out details there are a number of alternatives which we can skip at this point. In our review, this model serves primarily the role of 'control system of managers'; the stress is on the people, not on the control function.

This model tells for every manager on the chart

(a) a rank in hierarchie, automatically linked to the rate of priority of decisions,
(b) an approximate indication of his field of work.

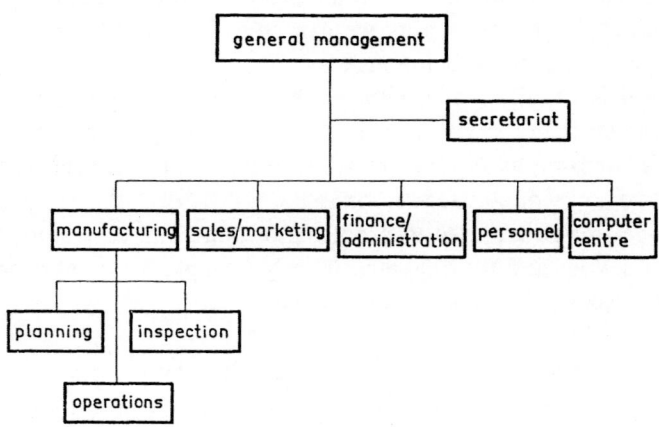

Fig. 5. Basic model of organisation structure

Since the chart techniques cannot contain sufficient details, the model usually is augmented with a set of non-formalised function- and task-descriptions.

In this system objects are the functions; we have to distinguish the functions (the chairs) and the people 'on the chairs', fulfilling the functions.

The systemrelations are hierarchy and communication pattern, both 'formal and informal'. Vertical lines in 'downward' direction have a 'command character' in addition; vertical lines in 'upward' direction have a 'report character' in addition. It is a well known fact that the actual (informal) situation may deviate from the official (formal) situation.

The Control System, described above, obviously is not a flowsystem. A clear and characteristic Input Output relation does not exist. This model cannot serve as communication system; the model does not explain the contents of communication flows, which rather are determined by other models, such as the technological models. Therefore we have to put this model of the control system into the group of structural (conditioning) systems.

*Ad* D

Procedures are flow systems. Objects can be data, forms, operations, data-objects (descriptors), data-subjects (participants)[5] etc. Relations can be time, sequence operations, authorizations (rights), etc,

---

[5] Participants in the organization here are considered to be all employees plus all external contacts: clients, suppliers, government, financial institutions, etc. We refer to Bouma, J. L., *Leerboek der Bedrijfseconomie* 1 A, p. 24 Den Haag 1968.

Procedures can be described with a variety of techniques. Many of them are charting techniques which show a flow in horizontal and/or vertical direction. Also other descriptive techniques are possible. E.g. sequence can be indicated by linesequence or by linenumbers or by matrix relations.

*Ad* E

Planning models (in technical sense) are mainly used for budgeting and for allocation of manufacturing, transport, finance and other capacities.

A planning model consists of the following elements:

– a procedure – based on certain suppositions – to select from alternatives a set of desired values (plan, standard, target) which values are considered to be a favourable or an optimal situation;
– a measuring system, which determines the actual values in a given situation;

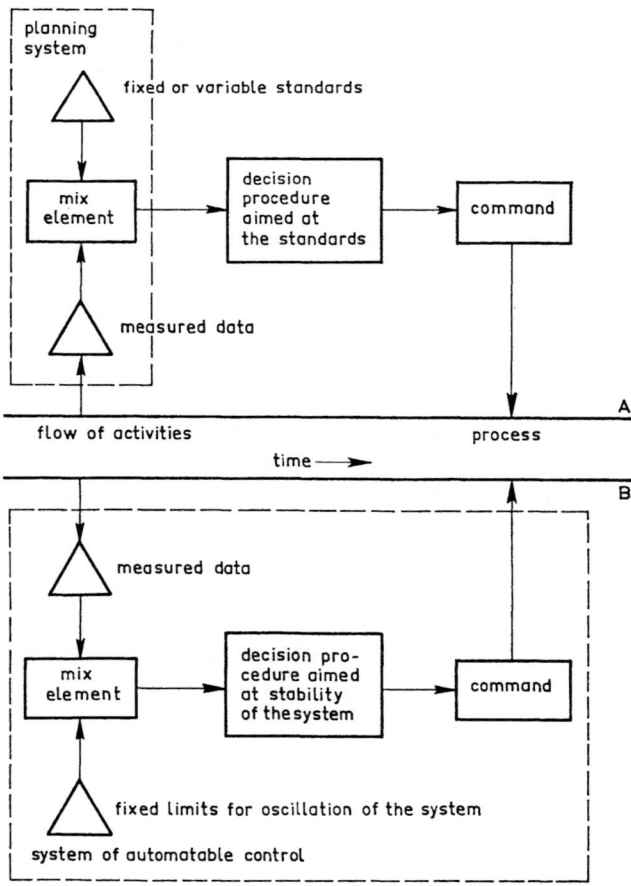

Fig. 6

– a calculation to determine any deviation between desired and actual situation; often this includes also relative deviations (e.g. percentages) and comparisons with other planning periods;
– sometimes a procedure to analyse causes of (substantial) deviations found.

The values considered are pertinent to a specific period of time: the planning period. The model of a planning system is a flow system, which processes 'data triples' (standard-actual value-deviation). Objects are these values; relations are said calculations. A planning model is given in Fig. 6A. The planning system processes data, which are used for the control (management) of a business firm or other organization, but the planning system itself is not a control system.

The essential difference between planning and control is that the latter contains a procedure resulting in a decision, as reaction on a found deviation; such decision procedure can be automatic. (It is a matter of terminology whether the results of such procedure then are called decisions or rather conclusions.) The objective of such decision is to bring about a change of situation in a way to make the output tend towards the standards[6].
A control model has been illustrated in Fig. 6B.

## 4. Evaluation of the above Models

All models listed above, can be considered as systems. None of them represents a complete description of a business firm. The question arises whether their combination makes a system that could represent a business firm in a sufficient degree of validity.

In our opinion it cannot, mainly for the following reasons:

a. Said models do not add up to a total system, a new unity: direct and logical relations between the various models are lacking.
The impact of a change in one model is not passed on automatically to the other models, but has to be implemented specifically.

b. The summation of said models does not produce a complete model of the business firm. This incompleteness is disturbing the full picture, in regard to lack of objectives, priorities and searching processes aiming for realization of objectives, in the models.

c. The descriptions of organizational structure and procedures usually only refer to the formal, prescribed situation; reality may deviate from such a picture, without notice.

Our conclusion is that the collection of said, traditional models together does

[6] Bosman, loc. cit. p. 220.

not constitute a valid complete model of the complex system which incorporates a business firm or other organization. Therefore, we designed a different approach through which a number of complementary systems together constitute a complete model.

## 5. A Model consisting of Aspect Systems

When looking at a complex phenomenon as a business firm, one often considers a specific aspect particularly, whether it be a financial side, a technical or a marketing performance, a management view, an employment profile or a point of development. According to our definition of system, any aspect considered can emanate a system, – Aspect System – which can be compared with a layer of the totality.

All aspect systems are part of the total; together they compose the total. Yet, aspect systems – or layer systems if you like – can be treated as systems by itself; the environments of such systems contain both its relations with the totality as its pertinent relations with the environment of the totality.

Our *general model* of an organization shows various (groups of) aspect systems:

– the technological and financial systems: all processes, which serve one or more objectives of the organization and which have to be controlled;
– the system of objectives and targets;
– the control system, in which we can distinguish;
– the control information system, containing decisions and their consequences;
– the control instrument or organizational structure;
– the procedures system, the level of processing of data.

All systems have levels of interrelationship and can be detailed at various levels.

Fig. 7 shows a general model of a business firm, based on aspect systems.

In our approach we choose as starting point the systems which have to be controlled, or the entire range of activities. The thought, that a business firm does not have one objective, but a range of objectives and targets, nowadays receives wider recognition. Objectives, including 'maximum profit' and 'continuity' have been enlisted in a range, with a rating for priority; the 'maximum' has been replaced by the concept of 'optimum'. Business aims can be placed in a systems hierarchy, ranging from the broad 'objectives' to detailed 'targets'. A specific technological or financial aspect system and specific objectives/targets show specific relations.

Control means influencing a system in such a way, that the system strives towards a given aim. Aims are a reference for the control. Control is initiated and maintained by decisions. The control system has two dimensions. First,

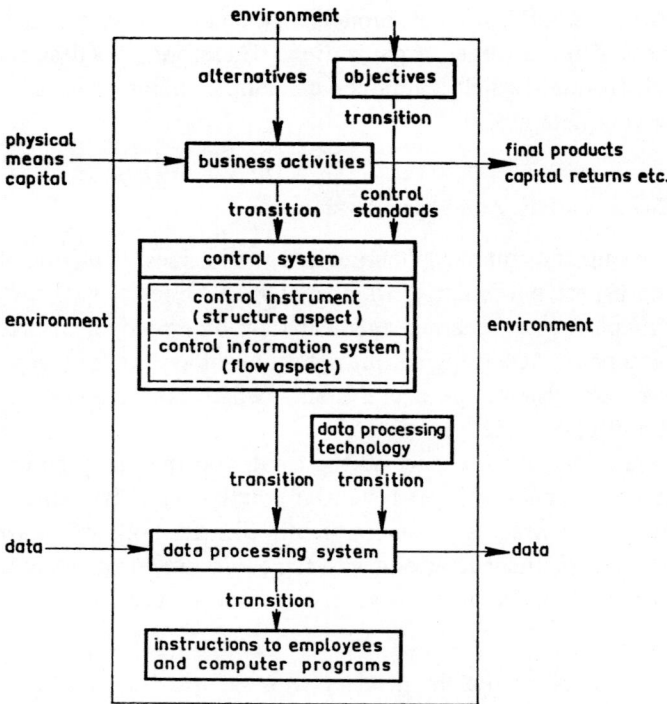

Fig. 7. Model of organisation with control objects contains the relations requiring formalization.

the control instrument tells how the decisions are initiated and made; its first level is the organizational structure. Second, all control is performed through information. Such information has specific functions, aims and usages; it has a pragmatic character. Its totality constitutes the control information system, sometimes called Management Information System (MIS) or Management Information Control System (MICS)[7].

All systems mentioned operate with the help of data. Data by itself are not interpreted and not linked to a specific usage; data have a semantic and a syntax character, but not a pragmatic one. The processing of data is described in the system of procedures; this system is a transition of the control information system and the data processing technology used.

The business activity systems (technological and financial), the control information system and the procedure (data processing) system possess the chaiacter of a flow system; outputs are identifiably determined by inputs. The remaining systems are conditioning systems. In fig. 8 the traditional partial descriptions are indicated.

[7] Maarschalk, C. G. D., The principle of organization Control Circuits, Progress Report in *Proceedings CIOS Congress*, Rotterdam 1966.

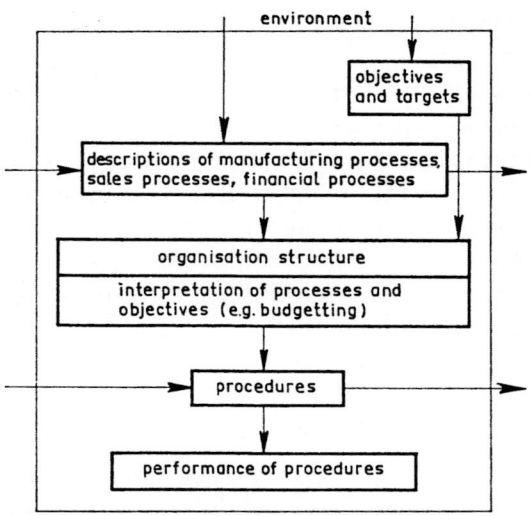

Fig. 8. Organisation model with traditional partial descriptions.

## 6. Example: The financial-Economic Model as Aspect System

In order to illustrate the position of an aspect system within the total system the financial model, given in fig. 4 has been completed and phrased as an aspect system in fig. 9. It illustrates that we can use the existing, well known building blocks but in a new arrangement adapted to system concepts and fitting also organization control purposes.

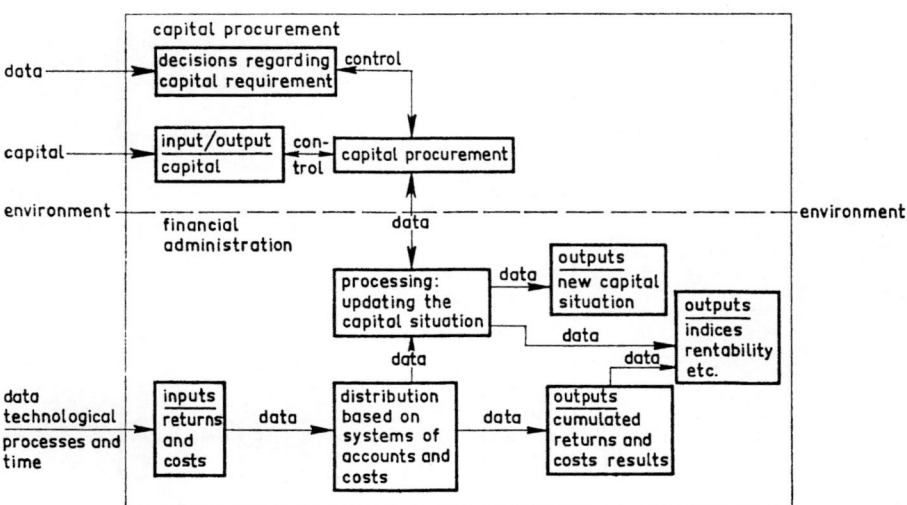

Fig. 9. Basic model of the financial aspect system

The elaborated model consists of two parts: investment/capital procurement and the calculation of results.

The first part is controlled separately based on capital requirements, costs, prognoses and capital policy decisions. Its output supplies data (capital and cost values) for the second part. The second part is not controled in a direct way, but through the control of other aspect systems, mainly technological models. The outputs (data) of technological systems become inputs (data) for the rentability model.

On short term basis the output of the latter is manipulated through the inputs of the technological systems. On long term basis there is an impact through the structure of technological systems and the output (capital costs) of the capital procurement system. This example does not claim completeness; it only demonstrates certain relationships between certain aspect models and its parts.

## 7. System Behaviour, System-Construction in relation to Decisions

The above illustration also serves as preliminary to the concepts systems behaviour and systems construction.

When reviewing systems behaviour, the output or the input-output relation is under consideration; the structure of the system is given. In the construction of systems, the structure is the variable. The structure determines a possible output at the condition of a certain input.

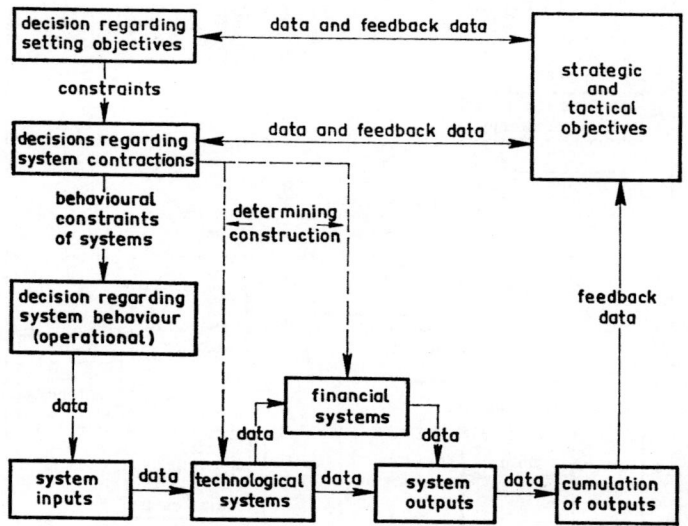

Fig. 10. Basic chart showing relations between kinds of decisions in a simplified businessmodel in which only technological and financial operative systems are shown.

In regard to control of systems behaviour, the input has to be manipulated in such a way that the output obtained is best possible in accordance with the given objective. The control of system construction has a wider range of alternatives, but has to take into account realisable system behaviour with objectives and restraints. It should not be overlooked that also the system of objectives/targets has to be controlled.

Therefore three kinds of decisions are required:

a. regarding system behaviour
b. regarding system construction
c. regarding setting objectives/targets

This is pictured in fig. 10. The latter group comes logically first. Construction decisions logically precede behaviour decisions. A change of structure should have an impact on behavioural control.

How does this all fit in the picture of the various aspect systems? A business firm requires a number of constructed systems, first of all objectives, technological and financial systems and control system. The construction of any system

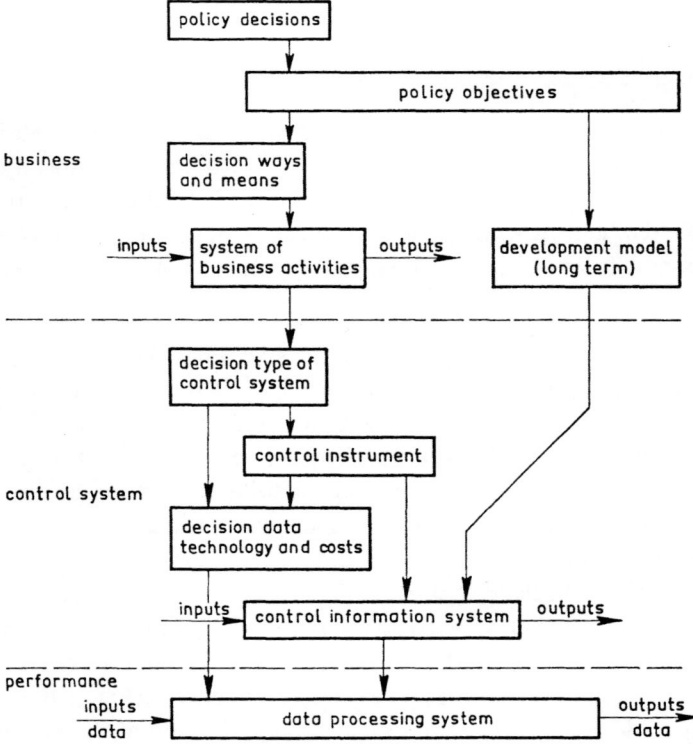

Fig. 11. Hierarchy of systemconstructions. The vertical arrows indicate directions of restraints.

cannot be isolated. Both objectives and restraints limit the alternatives; in fact, objectives and restraints have the same limitative effect. One aspect system can provide a restraint for another system; capital available is a clear restraint for a manufacturing system. The human aspect – or social system in our terminology – is a restraint in many other systems.

It is logical to start the range of constructions with the constraints which have the widest impact. Usually they are indicated as policy or strategy. In a business firm it concerns the selection of products, markets, capital procurement, social policy, development policy, etc. Policy variables are constraints for all system constructions. Further on, constructions of technological and financial systems, as well as objectives, become constraints for the control system. Consequently, a hierarchy of system construction emanates, charted in Fig. 11. It makes a model of the relations between the major aspect systems (layers), as to the system construction.

   This model contains three sets of flow systems: the business activities (technological and financial systems), control information system, and procedures. The outputs (the behaviour) of the first set, cause traffic in the other flowsystems.

## 8. The scope of Aspect Systems

The concept of aspect systems is meant to contribute to the development of systems theory and its application in the field of economic and social organizations. Further details can help with the analysis of organizational problems.

   In the field of general management problems – and management consulting – objectives and technological or financial systems can be considered as main variables.

In the field of automation said systems are given, but the control system and the procedure system then become main variables. In our experience, the concept of aspect systems has given a new opening for analysing the usage of a computer in organizational control. In order to place the computer as part of the control instrument and transfer to the computer a suitable part of the management function, further analysis of decision models was needed.

   Consequently, we developed several types of control systems; their applicability depends both on systems to be controlled and on policies.

   How do the traditional 'systems-terms' fit in our approach? The term 'systems design' means to us specifically: construction of the control system (control instrument and control information system) and construction of the procedure system, as to detailing. The term 'systems analysis' would fit the analysis of the given entities: technological/financial systems and objectives. Also we have listed specific qualitative requirements for the label *management information systems*.

We realize that this approach differs fundamentally from the view to apply a computer only for streamlined dataprocessing. However, when proceeding with the research on systems, in our opinion we are bound to find that logic in manmade organizational systems and logic in manmade machine systems somewhere meet. We hope that we have made a small contribution with this paper to that effect.

# SYSTEMS THEORY AND EVOLUTION

M. R. MANTZ

## Systems theory

The *Systeemgroep Nederland* would have no reason for its existence if the possibility of a generally applicable systems theory were doubted.

It is by no means sure that G.S.T., short for general systems theory, is such a theory. For the moment I can only conceive this G.S.T. as a set of generally valid principles and techniques which mainly are concerned with the technical aspects of labour processes and their relations. Examples are the black box approach and feedback systems.

The systems approach in the first place pays attention to the relations between the elements which compose the system.

In the notion of an open system one advances a step further. The relational pattern is extended with an interaction between the system and its environment. Precisely this interaction is the starting point for a further development which offers the possibility to develop and explain systems which are used in practice on the basis of an evolutionary process. In this picture interaction and evolution are inseparable.

*General properties of evolutionary processes*
*Change.*
It is evident that one cannot talk about evolution without change. However, what is causing that change? In my opinion one of the main motives is boredom. Already Spinoza pointed out that becoming aware of a change in the sense of being able to do more, is accompanied by a certain feeling of well-being.

From this the continuous creation of new ideas is explained. Application of these ideas results in a disturbance of the existing equilibrium of mutually adapted relations. This leads by nature to resistance. If nevertheless a breakthrough takes place, gradually increasing adaptational defects arise with respect to facets which have remained behind in developping, a phenomenon known as cultural lag. Remarkably the thus arising tensions begin to function as motives for a continuation of evolution. In that process the need for improvement shifts to the aspect that lags behind. This goes on until the tensions concerning the facets that lag behind have been reduced to such an extent that one may speak again of a harmonic equilibrium.

*Change in practice*
In our society we have not far to seek in order to observe these changes in our way of living, moving, educating, managing and automation. It is remarkable too that all these developments follow the same pattern. It starts with a technically creative stage, changing into a system of economical exploit with a tendency to enlargement of markets, while finally the cycle is closed by reflection on making subservient the acquired techniques and methods to man as he lives and works both as individual and as group in an organised system. A new attention is paid to the specific and the incidental on the level of the individual beside the uniform mass-handling approach.

This development can be observed in our educational systems. In spite of the fact that there are many means of improving the mass education e.g. by audio-visual means, more result is expected from a computer governed conversational system on individual base.

The invention of the steam-engine opened an era of uniform massproduction, of masstransport along railways. At the moment numerically governed machines make possible the automatic production of individual pieces of work. The automobile gives more individual service than the train.

*Evolution continuously changes its aspects*
Every development has a tendency to run on in its present direction. Technicians continue to improve their inventions. Economists continue to block further development because investment costs money and exploiting the existing brings in money. They only remain interested in enlarging returns and lowering prime-costs. Without central management to turn the centre of some development to a future aspect a one-sided development continues until it simply breaks off in want of interest from buyers.

Another phenomenon accompanying evolution is its accumulative effect. This means that a new system is more of an addition to the old system than a replacement of it. During development, systems are continuously better tuned to each other and by that become more vulnerable. If the higher system fails only the more primitive system is able to set the process going again and to correct the error. So in the new situation it functions as a stand-by. This latent presence leads e.g. to typical difficulties in the organisation of the firm where often an employee is not aware of the fact that more than one system of behavioural rules exists, each valid in a given situation. The leaders too are not always aware of this. We return to this question later.

*Characteristic changes in the organisation of the firm*
The oldest system is the pyramidal authority structure. The oldest relation is the personal authority relation. Through the rise of science and technics the functional authority relation originated, which is authority on basis of expert knowledge. In the utilization of this expert knowledge we notice a strong tendency to massproduction together with a strong division into more and more functional departments. The homogeneous character of a functional department however is undermined by the wish for increasing diversity of products. To comply with this wish a further differentiation is necessary which in its turn makes coordination difficult to achieve.

Then at some moment one chooses an organisation that leads the flow of the products along the productionline to a central starting point. This flow-promoting organisation, which beside the production of one given commodity may also be directed toward the serving of a certain market or category of clients gives the organisation again a number of qualities of a service nature. E.g. delivery times can be better predicted and maintained, responsibilities can be better defined. Communication is simplified and speeded up. The reactive power is reestablished.

The organisation splits into working companies who use the expert knowledge concentrated into the central departments. The relations inside the working companies become less formal and are more based on individual and to a less extent also global expert authority. There is more room for the development of enterprise. The relation to the central department resembles that of dealer and client. It rests on agreement resulting from consultation. The idea of authority hardly presents itself. The fact that an internal relation is characterized by the – already known – external relation dealer-client, indicates that working companies may enter also into such agreements with external experts and also that central departments may have external clients without changing the organisational system. Thus the boundaries of the system fade away and they become only manifest in the responsability for its results towards central management. It is remarkable that on the one hand fusion continually is leading to larger communities of interests whereas on the other hand simultaneously modules originate for obtaining a better flow. These are to be looked upon as satisfying supplementary conditions for integration into one system of economy and better service to producers and consumers.

## The sequence 'Technics-Economics-Service' in the automation of data processing

Here, i.e. in the automation of the processing of data, the aforementioned trend is even better observable than in the organisation of the firm. Originally designed as an automation for the performance of complex and lengthy calculations in operational research the computer has found its most extensive domain of operation in massive administrative works; because in order to evaluate the performance of a firm enormous quantities of data have to be processed into written reports and tables. On the base of these reports it can decided afterwards if certain standards of productivity were satisfied. This form of massprocessing could only be used globally for directing the goods flow in the form of periodic planning. In case of a break-down in consequence of some unforeseen often even unpredictable event the system had no solution and one had to rely on human ability to judge. Often humans are able to estimate troublesome situations very quickly and invent alternative solutions. However without sufficient information this remains a tricky business. Recent developments therefore are directed toward the construction of communication systems in combination with data banks, that are able to supply on the spot at any moment the required data about recent situations. Here again the techniques are made subservient to the continuation of flow of the production process and that by an increase of individual performance. This is achieved by giving on the spot the individual direct access to the remote central computer capacity. It is the same man-computer conversation that also in educational systems creates new possibilities for individually adapted education.

Space and time allotted to me are insufficient to introduce the problems adequately. Any discussion in this domain is bound to be incomplete, disputable and reflects the view one wishes to emphasize. In our case it is the wish to notice an evolutionary process. Besides, it is not the case that the three phases always occur after each other. In the stage of technical implementation of an idea, one is already concerned with its economical exploitation, and in the exploitation phase one is already wrestling from the beginning with the misadaptions, which are only satisfactory solved in the last stage. Besides every phase can be split into subphases which show the same structure. Evolution is hierarchically structured but moreover has a creation flow, which is characteristic for the properties of a process. Therefore evolution is merely a process with all its characteristic properties.

It is remarkable that evolution shows so much similarity to the process like structure which nowadays characterizes also the internal organisation. The technico-creative evolution stage has its pendant in the planning stage in organisation. The economic exploitation is comparable to the executive stage of plans, while the evolution toward subservience is nothing else than evaluation or verification. The analogy is striking but by some afterthought also evident. In both cases we have to do with cyclic labour processes.

In evolution this structural form is applied to use science and technique economically and to improve its usefulness later in an ever increasing domain

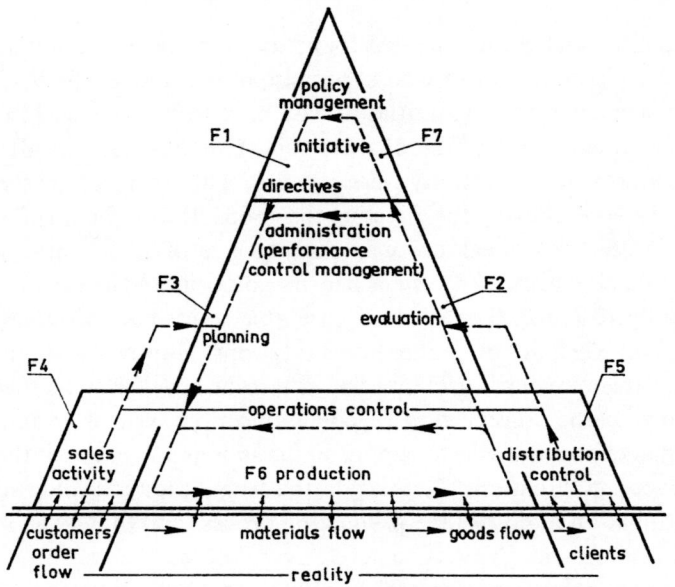

Fig. 1. Interaction between management of typical enterprise processes

## THREE LEVELS OF EVOLUTION
### (more levels are possible)

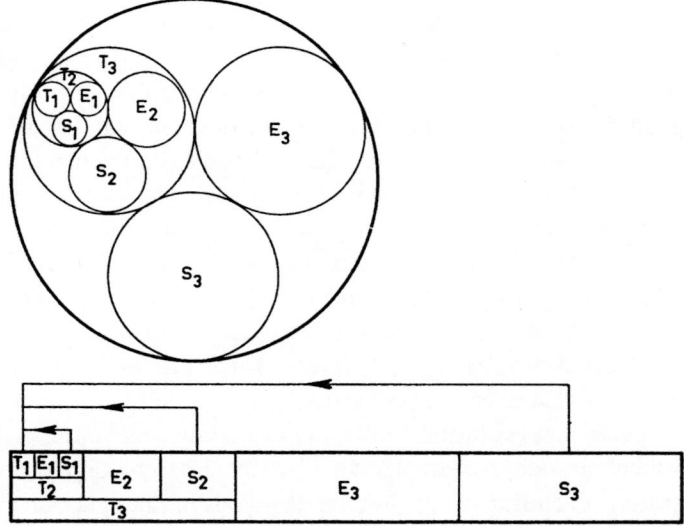

T = Technical – creative

E = Economic – exploitation

S = Evaluation to make the system subservient to man, management and society

T1 = Technical creative development for applied science

T2 = Technical creative development for economic exploitation

T3 = Technical creative development for making system subservient to man, management and society

Fig. 2. Evolution Cycles

## THREE LEVELS OF IMPLEMENTATION
(more are possible)

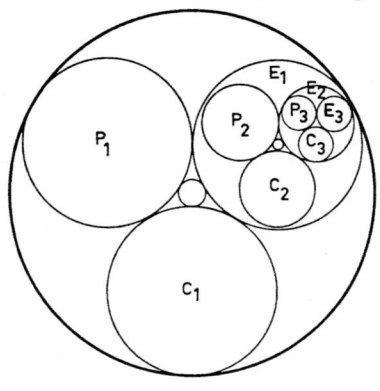

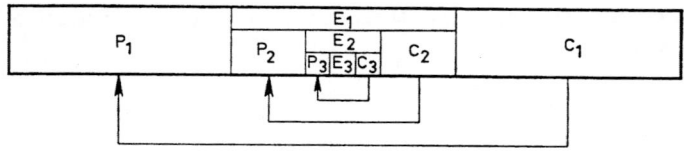

P = Planning

E = EXECUTION or
implementation

C = CONTROL or evaluation

P1 = Planning on strategical – innovation – policy level

P2 = Planning on tactical – implementation – performance control level

P3 = Planning on operations control level

Fig. 3. Economic Exploitation Cycles

of application (cf. fig. 1). In management (cf. fig. 2) the economical exploitation procedure is structured after the same rules.

In figure 3 of an elementary model of the firm that I often use to explain the stages of development of automation $P_1$ corresponds to managementplanning, $P_2$ to administrative planning and $P_3$ to operations planning. These stages are:

$F_1$: Operations Research

$F_2$: Administration

$F_3$: Productplanning

$F_4$: Orderacceptance

$F_5$: Distribution planning

$F_6$: Workshop scheduling

$F_7$: Simulation of the firm on strategic level

What meaning has the evolutionprinciple, c.q. the processstructure for a general systems theory?

I do not think that it gives a recipe for a model of the firm. It does however show the weaknesses of many empirical models. For it is possible immediately to recognize the evolution level and to predict the following level. The stage in the transition to a next step is easily recognized in the observed problems and tensions which are easily explained. In most cases one locates the reason for failing in personal leading qualities, which however is the way of least resistance to avoid the search for the true reasons.

In Table 1, the development phases of *automation in management processes* are presented in particular to illustrate in which respect applied computer methods can be made subservient to improve individual work conditions and management activities.

### The evolution of organizational structures viewed from a generalized process structure theory

The next few paragraphs, already written before this article was prepared, are added to demonstrate how, by assuming that an enterprise should be considered to be a pattern of processes, the weaknesses of well known organization structures can be explained. The 'line' organization has solved only the problem of labour division: Who does what. The 'staff and line' organization was only concerned with the problem how to use expert knowledge effectively. Both systems did not provide any means to create data flow and production flow other than by the wish to coordinate and to communicate.

When such systems become complex they turn into real bureaucracies where all departments do their work well but nevertheless render bad service to people in their work, to managers who should control and to clients who should be served promtply.

To promote the subservient character of an organization means in practice to design for fast process flow and better throughput what is precisely the challenge of today.

The triple symmetry indicated in fig 9 (pag. 56) can be found in practice, however the asymmetric development as mentioned above is the most current one and can be recognized after some practice. It has also become clear to me to distinguish sharply between process principles and economic principles. The integration of these principles always constitutes a second stage in an evolution cycle. The third stage then is the elimination of mis-adaptation which I have called development toward subservience.

| Phase | | Technique created | Economical exploitation | Subservient to | |
|---|---|---|---|---|---|
| | | | | Individual | Management |
| Technical Phase | T | Computer as computing instrument. | *Optimal solutions* for complex problems related to the lay-out of an enterprise. | Breaking through human limitations in computing power. | Improved policy-decisions. |
| Economic Exploitation Phase | Ea | Computer as automatic datahandling process. Extension with fast in/output equipment. | *Cheap mass processing* of administrative data | Monotoneous work taken over by machines. | Better management control of business performance. Also better production control. |
| | Eb | Computer as integrated *data process with short term around time.* Extension with direct access mass storage. | Improving the throughput of administration by individual processing, e.g. delivery of orders. | *Continuous* operating of the *workload* over the day. No occurrence of periodic peakloads. | Simplified administration able to give flexible service. |
| Service Improving Phase | S | Computer as instrument for *process control* by fast integrated data-processing. Extension with: · terminals · communication network · data base and also with · operating system · system software · compilers · problem oriented languages and: *standardisation* wherever it can be applied. | Improving the throughput of *management processes* by reduction resp. elimination of waiting times, stocks, buffers, etc. – Automation of house keeping system by system software and operating system simplifies greatly co-ordination activity in centre offering simultaneous service to many individual users. – Standard application packages safe time and money, speeding of greatly the implementation of automation. | Improving *individual productivity* in heuristic development work. Uninterrupted concentration on one problem. – Great freedom to use computer power where, whenever and as long as needed. Subservientness of computer to the individual man in his work and in many aspects of social life. | Improved performance of research; development- and project-departments. Faster and better financial reporting and control. – *Operational control* by direct information of actual situation. – Improved and faster decisions by man-computer dialogue. – Automatic registration of transactions solves data acquisition problems. |

*Principles, components and structural forms of the organisation*

*Components of the organisation*

To organize means to tune means and methods to objectives in such a way that the resulting relations can stand the test of efficiency, economy and reliability and the planned organisational structure is realizable.

As to the creation of new conditions, the work can be compared with the planning activity of an architect. Therefore it makes sense to talk about systemsarchitects and systemsarchitecture also in the process of automation of the organisation.

In case of a clear-cut task then the architect finds only support in his general knowledge, comparable to knowledge of raw materials, building-methods and principles of structure. In most cases he can rely on previous experience in similar projects. The greater this similarity the quicker and more purposeful his design will take shape. This is even more so if it is possible to use standard blocks of design, however the applicability of complete standard designs is limited. A solution therefore has to be sought in manipulating standard sub-units which equally permit the quick composition of efficient and economically well-designed solutions. Since it is easier to design standard units of limited application specialisation is to be expected.

### General organisational principles

The number of general principles which are not machine-bound and fundamental structural forms from which the systems architect can choose is relatively small. The starting-point is simple. A need of which one has become conscious can be formulated as a purpose. Any goal can only be reached by a process of labour, and man proceeds deliberately towards it, he first makes a plan, i.e. a realization of the goal as an idea. This plan is a clue for the execution. To be sure that the execution proceeds according to the plan, a control is necessary and this consists in comparing the real result with the planned one. This is quite elementary and well-known and leads to the distinction of three main functions:

– planning *P*
– execution *E*
– control *C*

Individual man can make more plans than he can execute. So he will try to delegate the executive stage to others, the more so because only if the plan has been defined one is able to formulate clearly his instructions. It is difficult to delegate what one cannot formulate. So the leader makes plans, the group carries them out and the result is checked by the leader (cf. fig. 5).

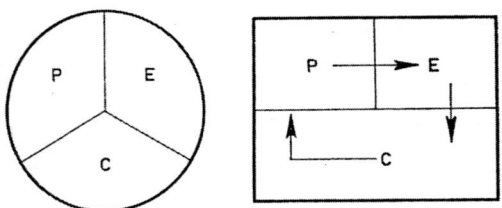

Fig. 4

If the group becomes too big so that the leader is no longer in a position to instruct and control its members, it can be structured hierarchically (cf. fig. 6).

The resulting organisational structure rests upon authority executed by the leader and accepted by the group. This structure is denoted by line-organisation

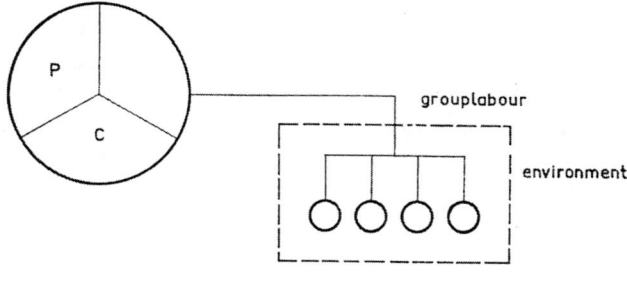

Fig. 5

and is one of the best known and most propagated organisational forms. The linestructure controls the relations between people in the group in charge of the executive part of the productionprocess. As has been pointed out the hierarchical structure is one of authority lines, it gives patterns of behaviour of man and knows the notions of superior, inferior, delegation of tasks, authority and responsibility. An important characteristic is the unity of command: every

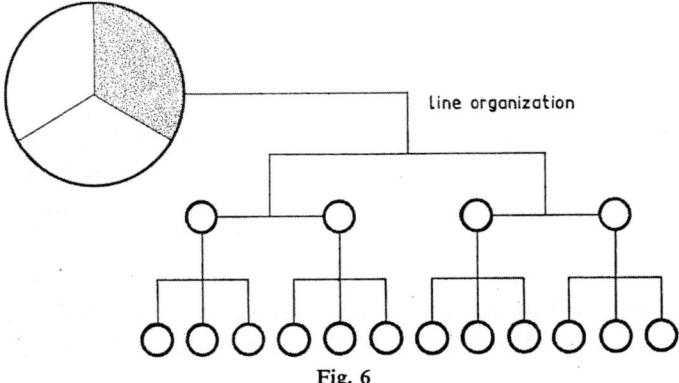

Fig. 6

subordinate receives its instructions from only one direct superior and from no one else. The subordinate is merely responsible to some superior. The pyramidal structure of the authority line pattern also warrants the possibility for two arbitrary workers to find one common superior who is able to settle controversies. This system of authority relations is completely adequate and cannot be improved.

However a factory does not exist for the benefit of authority relations but to produce things. Viewed from the point of a connected production process the pyramidal structure certainly is not an ideal one. This becomes evident if we let continue the leader to delegate certain functions to subordinates. It then will not remain restricted to executive labour. He also will wish to delegate planning operations to suitable experts. Planning operations require intellectual labour and the result will be a report or advice to the leader. In that case we say that the expert staff has advisory competence which really is more of a duty and this term is used in a restrictive sense, expressing the fact that it has *no* commandatory competence over workers in executive lines, although these workers execute the plans developed by the staff.

It is almost self-evident to continue in this direction and to delegate to the planning staff also a piece of control, with the same restriction, because it already knows the plan the results of which have to be tested. This structure is called staff-line-organisation (cf. fig, 7).

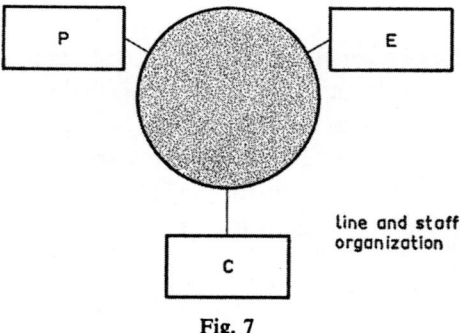

Fig. 7

All this is strictly according to the rules of delegation. The man in charge divides the tasks and receives the results. That is to say all communication runs via the central leader. As long as this is the case the unity of command is not in danger.

From the communicational point of view, the flow of information in the productionprocess however is far from satisfactory. The position of the already over-burdened leader becomes a bottle-neck in the communication system which starts operating cumbersome, slowly and inexpediently.

Direct contact between planning and executive departments carries the danger in it of cutting the unity of command. This can be a source of continuous conflict if rules for authority and functional communication are not sharply distinguished.

It is not always easy for a functionary to distinguish operational instructions from commands. In particular if the operational instruction is emotionally loaded with the appearance of a command. The problem is not solved however by psychological advice as one encounters often in the literature. Therefore we distinguish both aspects sharply although in practice the facts are not presented in such a black-white contrast.

The easiest way of characterising the authority relations is by: 'the boss knows, so there', Commands are not tested on their reasonableness but unconditionally obeyed. The other extreme is the operational advice, deduced from facts, expertly processed and based on logical consequences. This in contrast to decisions of the commanding authority, where personal insight, intuition and sometimes consciously risks dominate.

The difference between operational instruction and command is that between expertly drawn conclusion and responsible decision. This difference may also be illustrated by the operation of a computer.

A computer program only contains instructions of an operational character. No one will view the computer as a breaker of the unity of command, nor the systemsarchitect who designed the organisational scheme nor the programmer who wrote the program. It makes no difference if one and the same procedure is performed by computers or by man.

The program has a directive function, it has as it were the competence to perform the direction. In the same way a functional staff may perform a directive function, in so far that its directives and programs must be executed and the staff may have at the same time at its disposal means of control to be secure that this happens indeed. This is called functional competence to inform the line. This still has to be distinguished from authority over persons belonging to the line.

As long as the planning staff makes good plans that are accepted by the line and executed, the leader need not bother. A large part of the communication flow can take place directly between planning, executive and control units, whithout the necessity for the leader to intervene. The leader only has to intervene if something happens not according to the plans. This is called management by exception, where the leader delegates part of the communication flow to those directly concerned.

The task of the leader in a certain sense becomes latent and by that the whole hierarchically structured authority system.

The organisational diagram of any firm generally gives a clear picture of the authority relations which however give little information about how the daily work proceeds. The computer technology has given rise to the need of descriptions of the procedural structure. These are called flowdiagrams or organigrams, which exclusively give information about the jobs and say nothing about authority. The streamlines in an organigram can have different meanings. In information processing these are restricted to the indication of data or the sequence of operations. In management diagrams different streams like goods flow, money flow, etc. can be indicated. Eventually it will be necessary to develop methods to account for the relations, also in time, between the different flows diagrammatically.

### Relation between authority structure and operational structure

The question arises if there can be established a connection between authority structure and operational structure, This indeed appears to be the case. For although we have treated authority and operational task separately, in practice they are never completely separated. The superior will always beside his latent authority retain an active operational task in the form of immediate directive influence on the informationstream.

Management means not only giving instructions but also correcting deviations. This is only possible if this superior has some authority not only with respect to the operations but also with respect to the program and the measurements of the controlunits. The program is made by the planning department. Planning is a somewhat sloppy denotation for the drawing up of time- and

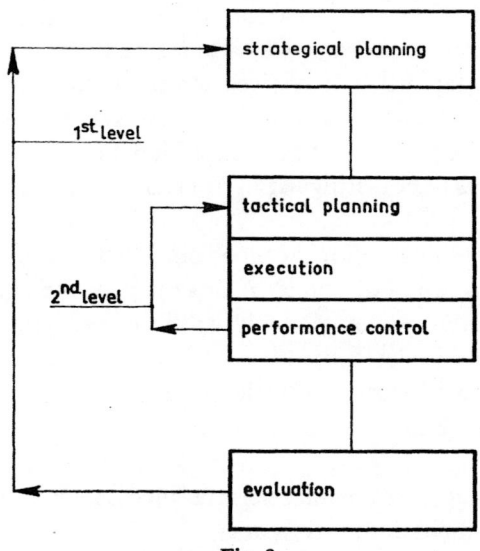

Fig. 8

occupation-diagrams and in general for everything that has something to do with preparation. Without a good preparation execution cannot proceed smoothly, and without control the preparation gropes in the dark. The three functions mentioned should be tuned closely to each other, which is a reason to place them as a threefold unit under the responsibility of one superior.

The preparation is preceded by a long-term planning. This planning which is more of a strategic nature needs also data about results obtained so far, however, in a different composition. This main planning is interested in sales figures and production figures over certain periods, generally in all kinds of diagrams and statistics to base a strategy upon for the future.

The control function that supplies these data to the main planning we could call productivity control. So we see a threefold unit on two levels, a strategic and a tactical level (cf. fig. 8).

The triplet planning-execution-control forms in a sense a closed loop of information which can be found on different levels, from global long term to detailed short term.

From the point of view of efficient management it is desirable that such a triplet as a whole belongs to the responsibility of one superior, and his conclusion leads to some structural rules for the combined authority and operational structure.

- Every labour process has at least a planning stage, an executive stage and almost always also a corresponding control stage.
- The control stage is not strictly necessary, but it is strongly desirable, because
- Execution without control is blind.
- Planning without feed-back by a control stage is fixed and sterile.

A central direction of this triplet is strongly desirable. This central direction laid in the hands of a leader is responsible for:
- the harmonic dimensioning of the three functions.
- a smooth flow of information with respect to the operations, hence for the communication between subsequent stages.

If a labour process is differentiated, i.e. that every stage becomes a separate process, the above mentioned triplet will again be found in the subprocesses. The subprocesses on a lower level have more the character of detail and require a different kind of leader. Obviously one should put a leader at the top of each triple, resulting in a hierarchy of the authority structure corresponding to that in levels of operational structure.

### Elementary structural forms

The connection between the elementary authority structure and that of the differentiated labour process have been represented in figure 9. Although

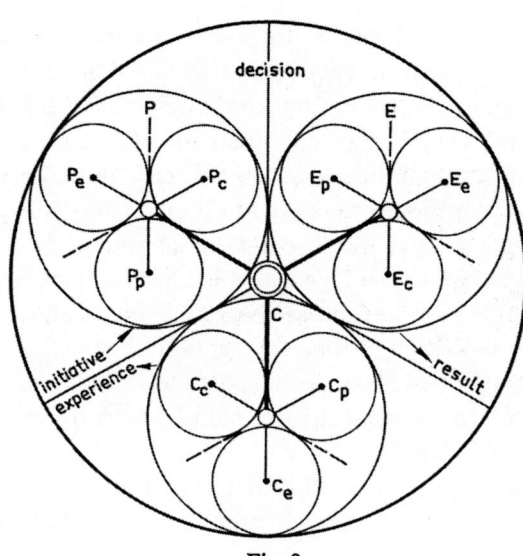

Fig. 9

apparently triply symmetric, each of the three functions $P$, $E$ and $C$ have typically a proper character.

In mass-production a given plan or design serves several times for the study of a productioncycle. The productioncycles have a plan creating stage. The controlcycle frequency will be intermediate between that of $P$ and that of the $E$-stage. Not every result requires testing and not every test lead to revision of design or plan. The production system has to be fed. Thus all planning stages appear to be meeting places of knowledge, respective tools for the subsequent executive stages. In general it is possible to distinguish:

– pre-processes which formulate the goals on basis of globally felt needs,
– post-processes for the distribution of results,
– auxiliary processes in support of the planning stage, in the form of special advices,
– auxiliary processes in support of the executive stage to supply materials, machines, personnel, etc.,
– communication processes in support of the goods flow and information,
– memory processes in support of following productioncycles, in the form of data files.

In figure 10 the above classification is visualized. It is a very simple structure pattern of an organisation. To this structure pattern belong a number of functions to keep the production-apparatus in good condition, for example a technical service to keep the machines in working condition, a personel department, a salary administration, a safety department, a building department, etc.

The functional competences belonging to a sub-, auxiliary- or side-process can be deduced from the role they play in the management. To this a second criterion is added, viz. it is expected that every activity is useful in some sense, so that the invested labour, time, knowledge and means have some effect. The actual interest in a good system of production regulation goes in this direction particularly in plants that know a high rate of mechanisation or automation. Formerly the increase in production was a question of more workers, today it is a question of more and more efficient machines. Until recent times the increase in productivity was a question of stimulating human labour by better systems of tarifs, today it is a question of better designed occupational diagrams, minimalising regrouping of labour force and timely reaction on unforeseen set-backs; all this is only possible by quick and timely information to improve the flow of production.

All things that have been discussed so far are elementary, far from complete

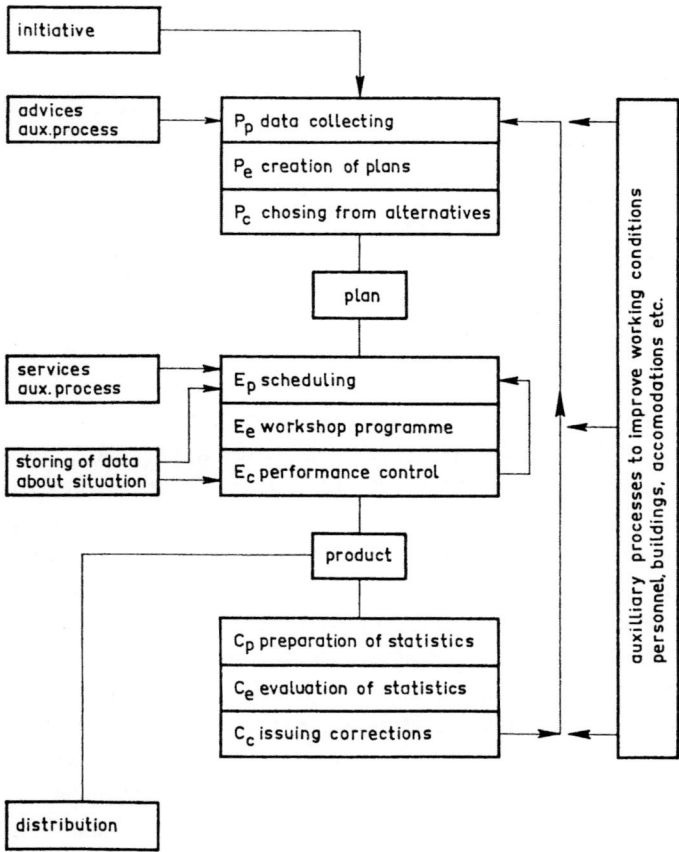

Fig. 10 Process classification

and formulated in a too general way to be of direct use in concrete situations. Nonetheless a general orientation as given here cannot be dispensed with in practice.

The organisational structure as depicted in figure 10 suffices if the objects to be produced differ very little. If this is not the case, the advantages are lost because of lack of homogeneity and conflicts arise about priority of the use of tools and rearrangement time and waiting-times occur as well as intermediate stocks. A functional department, i.e. one specialized in one kind of operation is mainly interested in a regular full employment of its capacity and not in a consecutive cycle of one type of product. The partial interest is given priority for the integral interest, and a continuous productioncycle is reached by revising

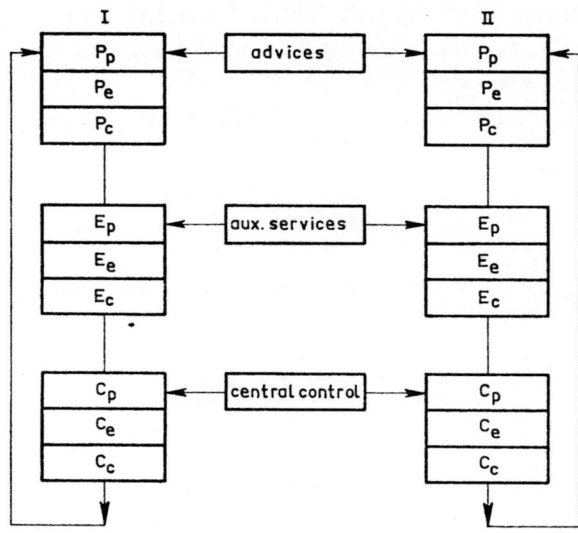

Fig. 11

the organisation in such a way that the production-apparatus is homogeneous for each produce. Then however the solution is of the kind exemplified in figure 11, and the specialization thus obtained reduces the production apparatus to a set of more or less independent factories each specialized on one type of produce. Although in that way a short performance time of the flowcycle per produce is obtained it is a pity that the separate production lines cannot profit from each others experience or technical provisions. It is a failure to leave the necessary interplay to a spontaneous cooperation and teamspirit.

An a priori appeal on cooperation and teamspirit is as unfortunate as an occupation plan that can only be satisfied by planned overwork. For this means that reserves are not available in unforeseen circumstances.

An efficient cooperation between related production lines should not be left to spontaneous cooperation but on the contrary be consciously organized.

This can be done by not separating all services, but using a number of staff and auxiliary services in common. This is called parallelization, bringing about a connection between independent production lines, well to be distinguished from integration, which reinforces the connection between the functions of one and the same production line. Integration is opposite to differentiation, parallelization is the opposite of production specialization. Particularly in automatic information processing parallelization occurs frequently. In microform in the production of copies of document, in macroform by concentrating the information processing into one centre to obtain a fuller employment of machines having a large processing capacity. In stead of integration and parallelization the terms vertical and horizontal integration are used. These terms are derived from a customary representation in diagrams and should be avoided because of their informational weakness.

*The economical aspect*
Up to now our starting point was the search for methods to realize certain aims. In society it is also required that the goals are reached with effective use of tools and materials. Competition and scarcity of means are compelling. Evidently the satisfaction of both operational and economic exigencies, require a higher level of organisational thinking than simply satisfying technical ones. A new technology at first is mainly technically oriented and only in a riper stage concerns itself with economical aspects. The neglect of the economical side of problems has in the past led to bitter disappointments, in particular in those cases where in the introduction of new systems the possibility of cooperation was discarded with existing systems in which great amounts of money had been invested. Organization has to be considered as a process of growth in which sudden changes have to be restricted in number and size. It is a gradual striving to obtain beside technical perfection an economical optimum and finally serves man, enterprise and mankind.

Each organizational arrangement has to be tested as to three principles, where the first one is of operational nature and the other two are oriented towards the economical.

Condition 1: Tuning of available knowledge, labour force, tools, time and materials to the aims set.
Condition 2: Minimal waste, i.e. no use or consumption of knowledge, labour force, tools, time and materials that does not contribute to the desired results.
Condition 3: Optimal use of invested knowledge and tools. This means divulgation of knowledge and full employment of machines.

These conditions are fundamental and universally valid. Every technical construction for example can be judged by these conditions. Besides, it appears that technology and organization have to satisfy the same fundamental requirements. The distinction between technology and organization is analogous to the distinction between hardware and software in computers. Moreover the above conditions appear to hold in dimensions of matter, space and time, to wit

- for matter as to quality and quantity
- for space as to place and size
- for time as to moment and duration.

The pragmatic organizer will ask 'What use have these principles?' For they give no presciption for the solution in practice. They are not meant to do so. However he who has to go new paths, as is the case in automation, he who has to think about solutions in the long run will, by lack of experience, find support in the products of intuition and inspiration.

However it is possible to give a slightly more detailed representation of the above principles as the next table 2 shows.

Table 2

| Principles<br>Dimensions | 1<br>*Tuning of means on goals* | 2<br>*Minimal waste* | 3<br>*Optimal usage* |
|---|---|---|---|
| Matter | Qualitative | Simplicity | Of qualities<br>e.g. know-how |
| | Quantitative | Economy | Of quantity |
| Space | On right place | Concentration in space | Of local conditions |
| | Of right dimension | Minimal size | Of dimensions |
| Time | On right moment<br>On right time-duration | Well defined moments/Minimal duration e.i. fast | Of periodicity<br>Of continuity and homogeneity |

Explanatory notes to column 2: minimal waste.

*Simplicity*
One of the major goals of both the machinebuilder and the organisation designer. Complexity often is a sign of immaturity of unsufficiently meditated technology. A mature technology in general is marked by the application of general principles of great simplicity. Even if the latter notion may not be objectively defined. The reason why at the moment one struggles with the

problems of primary fixation in machinecode is because existing methods are too complicated and expensive. Sometimes one succeeds in avoiding a problem, e.g. the input-output problem for large data collections is partially avoided by storing them in for the computer easily accessible large-scale stores. Often it is not necessary to print mutations and only the passing of certain boundaries is of interest (management by exception). This gives a partial elimination of dataoutput. Further automatization probably will make it possible to create many basic data completely automatically without human interference. Another example induced by the requirement of simplicity is restriction of types, while simplifications resulting from standardization give more profit than the loss of a small misadaption.

*Economy*
No comment.

*Concentration in space*
Important in this connection is the avoidance of unnecessary transport of goods and data. E.g. data transmission easily spans time and distances however if a computer can be installed at the site where the data are used one eliminates the transmission problem. A firm that does not install its computer at the central storehouse thereby creates a communication problem.

*Minimal size*
In the technology everyone is striving towards smaller size. This gives direct savings on material, storing space, transportation costs and installation costs.

*Well defined moments* (synchronization and coordination).
Synchronization is important if a number of operations have to be performed in sequence. In case of equal operation-time and uniform transport it is possible to keep all units occupied. Therefore combinations of punchcard machines operate at uniform speed.

Coordination is important if a number of operations take place simultaneously, the results of which have to be combined in the right moment. This is one of the main problems in project planning. Synchronization and coordination are of great importance for the regulating of production and distribution. If all products would be ready at the moment they were needed, stocks could be cleared away.

*Minimal duration*
The economical use of electronic equipment in administration is based on its high processing speed which enables an enormous production. A relatively low price combined with an appreciable degree of occupation will result in a low

price per operational unit. Since in mass production initial costs are negligeable there is a tendency for the price of a tool to be determined by the quantity of material needed, which in view of the small size of electronic switching circuits is very low. So it is to be expected that eventually the price of electronic equipment will fall. Economical effect in general is only arrived at by a combination of factors, which as far as can be seen have appeared to fit into the above table.

Explanatory notes to column 3: optimal use (of equipment already present).

*Quality*
The notion quality has to be defined as the knowledge necessary to yield a required performance, with the inclusion of necessary equipment, accomodation, in short everything that has to be present, has to be invested, before production can start. Quality is used in the productionprocess and not consumed. Knowledge e.g. can become obsolete, but after a production cycle it is undiminishedly present. Capital wears and becomes antiquated but in principle is again available for the next production cycle. The acquiring of quality like knowledge costs time and money. The more this knowledge can be used the higher the yield of the costs of investment. The investment in intellectual capital in automation is extremely high and have to be compensated by intensive use. Uninterrupted use of computers – if possible 24 hours a day – therefore is a natural striving, provided the problems of operating and guard can be solved, perhaps by further automation.
Standardization is a means of crystallizing acquired knowledge and can be used universally. For because of standardization it is not necessary to develop anew approximately the same system, program or machine.

*Quantity*
This has been treated in the preceding section. By increasing effective productiontime and the quantity produced per invested quality.

*Size*
The advantage of small size is mainly that of easy handling, cheap implementation, requiring little space, low costs of materials. Often small sizes contribute to a lower influence of inertia of moving parts, resulting also in higher speed and less wear. For information carriers this implies lower transportation costs, lower storing costs and lower acquisition costs.

*Periodicity*
Without periodicity a continued occupation of capital goods is hardly realizable. Regularity causes planning activity to be simple and therefore less expensive. It is dynamic if viewed from the same point as simplicity statically viewed.

*Continuity and homogeneity*

Continuity is a very important principle. Idleness is loss and reduces economical effectiveness.

Homogeneity is desirable because a variety of work requires rearrangement of the production. The time for this is unproductive and reduces effectiveness. Division of labour and specialization use homogeneity to obtain continuity. Homogeneity means that knowledge and capital goods can be restricted to that necessary for a single type of operation. In case of diversity of work always a part of the available means stands idle, which already means loss in effectivity. On the human level things are different. A diversity in knowledge is naturally present and change seems to be a natural labour condition. Indeed the importance of the economical aspect should not be overrated and other aspects are not to be forgotten. In the practical organisation we have already mentioned the test on realizability. As a fourth criterion we could mention reliability. Increased reliability may be obtained for example by duplicating storage, processing means, recomputing of results arriving along independent paths. Duplication of provisions is not always necessary although a certain redundancy is unavoidable.

The economical principles for technical and organizational construction have been applied as far as can be remembered. From the examples it can be seen that generally a combination of principles is necessary.

Speed in itself does not suffice, but has to be combined with homogeneity and continuity. A number of such combinations are already patent in classical theory of organization like specialization (according to produce), differentiation (according to functions) and standardization.

Economical arrangements always have a certain strained character, they often are detrimental to other qualities. Specializing detracts from simplicity of coordination, differentiation detracts from the integral character of the organization, the advantages of a decentralized commercial efficiency are sometimes hindered by lack of central coordination. As we shall see presently one of the great advantages of electronic switching- and transmission-techniques is the possibility to integrate some older compromise arranging. To see this one has to recognize that specialization has to be balanced by coordination, differentiation by integration, decentralization by centralization, routine by improvisation and standardization by individual tuning.

One cannot enough emphasize that the pairs of notions are not pairs of opposites, but of necessary complements. It is curious to observe that the respective first mentioned conditions in the large firm have contributed so much to the increase of productivity, whereas the second conditions may mean the effective power of the small firm, because there they are easiest realizable.

# CONSTRUCTION OF A COURSE IN
# TECHNICAL MECHANICS: A SYSTEMS APPROACH

W. MEUWESE

In education complex environments are created to reach complex objectives. It therefore pre-eminently seems to be an area where systematic thinking and decision making can be supported by the so-called 'systems approach'. This means that the process of construction and evaluation of educational systems perhaps obtains a more rational and systematic character, if this process is guided by concepts and methods of systems technology. Application of this conceptual apparatus to education as yet will have a very informal character, however.

As an example, this paper describes a project of construction and evaluation of a course in Technical Mechanics for freshmen in Mechanical Engineering at the Technological University Eindhoven[1].

Starting point for a system construction project is the assignment to realize a desired result within given constraints. To that end, this desired result is analyzed in criteria adequate for the given problem area (the desired output or the system goals). Next one tries to regulate the system input in such a way that the desired output is obtained, without violation of the constraints. In essence this means, that the system is built in such a way, that the course of the output is predictable from the input.

In such a project knowledge of relationships and processes within the system is important only insofar as it adds to improvement of the input – output relationships. From a system-technological viewpoint the better the output can be regulated by variation in input, the less important this knowledge is.

System construction and evaluation contain elements of classical educational and psychological research. For example, measurements should be reliable and valid, therefore measuring instruments will be constructed on the basis of the usual psychometric models. In the system construction process theoretical

---

[1] Project conducted jointly by the Technical Mechanics Group of the Mechanical Engineering Department (W. Esmeyer and L. Braak) and the Educational Research Group (W. Meuwese and H. Tielens).

knowledge obtained from basic educational and psychological research is used. But there are also differences.

First, measuring instruments aie not constructed as specifications of abstract theoretical concepts, such as anxiety and spatial ability, but derived directly from the system goals. Second, the system manipulations – which are comparable to the experimental conditions in conventional research – are not systematically varied, but iteratively optimized. Third, the system outputs are compared with a norm – the desired output – and the complete system is evaluated as to utility: benefit versus cost. And finally, research on the system is primarily oriented toward improvement of the system itself.

As a secondary research goal it is stated, however, that the results should be generalizable to the extent that they may contribute to general knowledge about methods for system construction.

**The construction phase**

The decision to construct an entirely new system for Technical Mechanics was the result of a two year period of regular measuremt of course output with multiple choice tests, and discussion about the results of these tests, about preferable objectives for Technical Mechanics in relation to objectives of the Mechanical Engineering Curriculum, as well as study of experiments conducted elsewhere.

The object the team finally formulated for itself was to realize an educational system for instruction in Technical Mechanics for students in Mechanical Engineering. The system should be adaptive to differences in learning speed, usable for approximately 200 students at the same time, and flexible concerning time allocation and study methods of students. Constraints were: staff was available only six hours weekly, technological media could be used to a very limited extent only, and the system should be operational for a first try in one year. Moreover it should provide data which could be used to evaluate utility and could serve as a base for improvement,

The system was constructed on the basis of a series of considerations, some educational, some practical, for example:

1. The objectives of the system are to be analyzed in advance. To this end Bloom's (1956) taxonomy of cognitive objectives was used.
2. Written text and experiments with mechanical constructions are the major teaching media. Written text because of the limited availability of staff, and experiments because of the emphasis on the relationship between mechanical engineering and theoretical mechanics in the formulation of objectives.
3. Learning process probably will be enhanced, if the student works indepen-

dently part of the time, when he follows a problem-oriented strategy, when he sets short-time goals for himself, and when he gets feedback on his progress.

4. Staff activities, such as scoring and interpretation of tests and giving advice to students should be structured in such a way, that automation is possible.

For practical reasons a 'self-paced' study course, in elementary physics, developed at the Massachusetts Institute of Technology, was used as an example to work from (Green, 1969).

### Paradigm for an optimal system

A paradigm for the construction of a system which was predicted to be optimal within the constraints given indicated as a first step derivation of hierarchically ordered clusters of sub-objectives from the total set of objectives. Each cluster then defines an instructional unit. Each unit consists of four parts:

1. A review and the formulation of objectives for that unit in operational terms. These serve as 'advance organizers' (Ausubel, 1968).
2. A number of suggestions for strategies of study to reach these objectives. Examples of possible suggestions are: a theoretical review is given in Chapter x of book A, practical applications are described in Chapter y of book B, C presents in article c a critical note. On mimeographed sheet b you will find a more workable symbol system. You can test the laws of mechanics described in this unit with experiments k, l and m. In room Q you will find a model of a bridge, which you can use if you feel like testing some hypotheses on real-life constructions. Problem series d and e, programmed instruction sequence f, and CAI program g can be used for practice in problem solving. ALGOL-program h can be used to process your data, perhaps after some modification. Concept p is illustrated on filmloop P, *et cetera*. Imaginative selection from these possibilities may result in an interesting and motivating multi-media system.
3. A number of rather complicated 'study questions'. Each question poses a problem that can be solved with the knowledge obtained in the course so far. These questions should activate thinking on a higher level.
4. A number of diagnostic tests, for feedback and forward and backward steering purposes,

After the course has been automated, the diagnostic testing and advice procedure can be done on a computer terminal. The terminal gives immediate feedback of results, and eventually steering-signals to specific parts of the unit if the result on the test is below a prespecified norm. The computing system can process data internally and provide the system-managers with tabulated results.

**First experimental realization**

For the first experiment with this system twelve clusters of objectives were defined, hence twelve units were put together. In the first try, which of necessity was only a meager derivative of the paradigm, each unit consisted of the stated objectives, a list of references to specific pages in books which could be used, supplementary texts, a series of study questions, answers to these questions, and six diagnostic multiple choice tests of approximately twelve items each. The medium in this first approximation of the expected optimum were for practical reasons restricted to written texts. The diagnostic tests were partly overlapping, to make it possible to project the scores on a common scale for each unit.

In the first semester of the academic year 1970–1971 this system was tried out with 180 freshmen in Mechanical Engineering. After an introduction to inform the students about the system and to ask their cooperation for the evaluation procedures, all students were given the first unit. At the time they felt they had mastered the unit they could do a unit test. This test was randomly chosen for each student from the six tests for that unit. If the score was below the norm, advice was given about material to be studied again. This advice was strictly on the basis of item responses, and was given by a graduate assistant who selected adequate advisory statements from a list of possible statements, following a specific prescription.

After a period of study the student could do a second test, and the procedure was repeated. If results at the third try were still below the norm, than the student was tutored by the professor. Two afternoons a week were available for testing. For each group of 15 students an assistant was available, who scored tests, monitored the advice procedure, and distributed materials.

**System evaluation**

This educational system can be seen as a laboratory situation. Evaluation of the system should provide data which can be used to improve the system, and should allow generalizable conclusions. The evaluation therefore is rather detailed, and measurements have to be expressed in terms of variables on an abstract level.

The *input* of the system can be divided into two classes: student-parameters and characteristics of the units.

On the first day of the program all students were tested. The battery of tests was chosen to comprise factors expected to be related to the learning process in this specific system, either as a direct predictor or as a moderator. Some examples are: intelligence, nonverbal abstraction, computational skill, symbol transformation skill, divergent thinking, achievement motivation, neuroticism and introversion – extraversion.

Unit parameters should be defined in terms of relevant dimensions of the learning tasks. Suggestions can be obtained from the literature on problem solving, concept formation and programmed instruction. Some possible dimensions are: redundancy, internal structure, implied variety of strategy, level of abstraction, complexity. To quantify these dimensions some objective indices can be invented, such as the ratio of mathematical expressions to text, the number of suggestions in the unit. However, judgments of experts on the course content carry most weight. Of course the reliability of these judgments has to be checked.

The most important measurements of *process* are scores on the diagnostic tests. Other indices are: speed of advance through the system, number of test repetitions, kind of advice given, and responses to questionnaires about strategy followed, time used, evaluation of difficulty and conceptual and structural clarity of each unit, and motivational factors. These questionnaires were given after each unit.

The *output* is measured by a multiple choice test. These data can be used to compare the new system with the conventional lecture – exercise system used before this one was introduced. This comparison is only partly possible, because the new system had partly new objectives, and filling out a test after one has passed the course is motivationally different from taking an examination.

Data can be analyzed from different points of view. Seen from the viewpoint of the system constructor relationships between structural characteristics, i.c. the unit parameters, and processes in the system seem to be of primary importance. These relationships can be investigated for the system as a whole, or for any definable subsystem.

Another point of view pertains to the predictability of process and output from individual parameters, such as testscores. This analysis provides data on ability and personality factors conducive to task success, on sequential dependencies within the system, and on predictability of success in an early stage.

A third point of view implies analysis of content. This analysis may point to parts of the system which are not sufficiently successful. To this end achievement test responses can be analyzed in detail. To make this analysis powerful, it may be necessary to structure the set of test items in such a way, that a mapping obtains of both the conceptual structure of Technical Mechanics and levels of intellectual functioning.

### First results[2]

A first impression of the system's output may be gained from Figure 1. In

[2] These results certainly are not definitive, and therefore should not be quoted.

this graph distributions of students over units passed are given in intervals of two weeks.

After 15 weeks, the time when the system was closed, an output/input ratio of .87 was reached. For comparison: this ratio for the conventional system varies from .50 to .55.

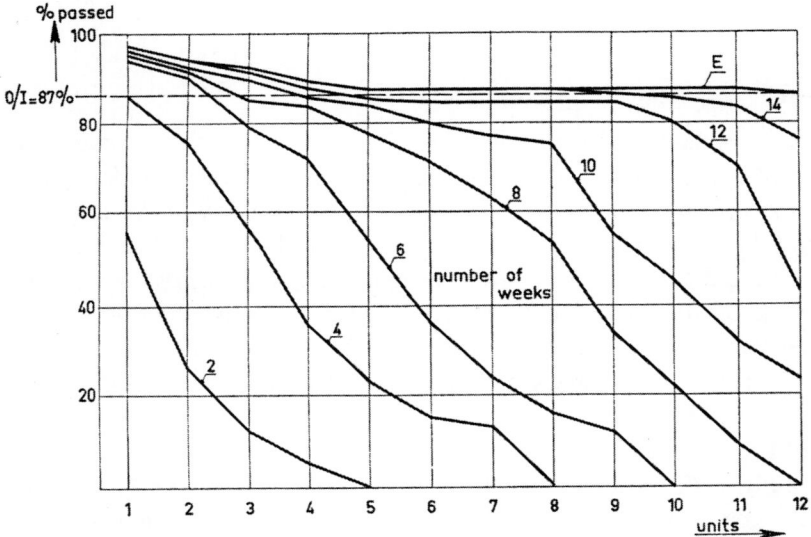

Fig. 1. Distribution of the student group over the units in intervals of 2 weeks

The spread of students over units is very large, especially halfway the course. The self-pacing character seems to be justified by this result.

Analysis of the criterion test, equated on examinations of January and July 1970, produced a lower limit estimate of the pass/fail ratio of .64, and an upper limit estimate of .75. This means, that a proportion somewhere between 64 % and 75 % would have passed if the test would have been an examination.

Some subjective reactions of the students can be obtained from a questionnaire given at the end of the course. The course was found pleasant or very pleasant by 74 %, captivating or very captivating by 78 % of the student-group. Fifty-seven percent found it worthwhile to try to structure most of the first-year curriculum in this way, while 19 % was neutral on this point. The multiple choice testing was found more pleasant or much more pleasant than written or oral tests by 66 %.

One can ask for the influence of this system on the educational environment. Questionnaire responses indicate that no disturbance has been experienced in the study for other courses. More objective results perhaps are examination results for mathematics which were on the average not different from earlier years, and the fact that more students than usual subscribed to an examination in Metallurgy.

The high output ratio of this system is not a result of a higher investment of student time. An estimate of total time needed to pass the course, obtained from the unit questionnaires, produces an average of 70 hours, 40 hours study and 30 hours testing. A lower limit estimate for the conventiona' system is 100 hours, an upper limit estimate is 165 hours, on the average.

## Next steps

The use of graduate assistants could only be justified because the course was experimental. Automation of test and advice procedures therefore is logically the next step. Automation on the terminal seems to be fairly easy, because the assistants have taken only a very low number of decisions which were not programmed. Automation moreover has the advantage, that system monitoring, testing and feedback can be more continuous, because the restriction to two afternoons a week is no longer necessary. The course material shall partly have to be rewritten on the basis of data obtained. Also some new media and instructional strategies can be introduced as a next step toward the paradigm assumed to be optimal.

Hopefully working along these lines provides sufficient experience to enable relatively fast construction of other courses based on the aforementioned paradigm. Already now it has been demonstrated, however, that it is possible to construct within a reasonable preparation time instructional systems which do not require a great labor force, which obtain a satisfactory high yield and which can be evaluated in a rational way.

## References

Ausubel, D. P., *Educational psychology, a cognitive view*. New York (N.Y.) 1968.
Bloom, B. S. (ed.), *Taxonomy of educational objectives I*. New York (N.Y.) 1956.
Green, B. A., *A self-paced course in Freshman Physics*. Cambridge (Mass.) 1969.

# SOME METHODOLOGICAL ASPECTS
# OF A SYSTEM THEORY

B. VAN ROOTSELAAR

## Introduction

The purpose of this lecture is two-fold. In the first part the role of the mathematical notion of a functional is emphasized and it is shown to be the underlying notion in straightforward practice in more or less formal systems constructions. In particular it is argued that in many cases the simple conception of a system as a set of input-output relations is inadequate.

In the second part the analysis of current notions of tension and a subsequent model construction based upon such an analysis is suggested as an object of study for some subgroup of the Systeemgroep Nederland.

## Functionals

In considering methodological aspects of a systems theory we investigate the essential properties of some notion of system and the possibility of successful application of such a notion in scientific research. One therefore tries to define such a notion with the greatest precision and general systems theory has in mind a mathematical description, preferably one using the mathematical formalism of set theory.

Besides a precise description one also is interested in obtaining the greatest possible generality. The endpoint on the way to greater generality is easily indicated. One of the most general definitions that have been put forward in the current literature is the following:

a system is a relation $S \subset X \times Y$,

this has to be read as a set $S$ of pairs $(x, y)$, where $x$ belongs to $X$ and $y$ to $Y$, and one adds that $x$ stands for input and $y$ for output.

If we look for generality it is clear that we can go one step further and simply define: a system is a set, and we have arrived at a complete triviality. It may be obvious that it is not generality we are in need of. We should look for a useful notion that is adequate to the problems we wish to handle. We therefore ask whether our first notion is adequate, e.g. is a coffee-machine a system in the above sense? That means, is it the direct product of a set of coins and a set of cups of coffee, i.e. a set of pairs consisting of a coin and a cup of coffee?

I would not call such a description adequate, for, besides the fact that it does not seem realistic it has other defects. The number of pairs is not defined, and if defined of little practical consequence, moreover the description does not distinguish between the different pairs.

It seems that we are better off if we give a functional description of the machine, e.g. by specifying

> if $x = p$ then $C(x) = c$
> if $x \neq p$ then $C(x)$ is not defined.

On this description of the coffee-machine, $c$ stands for a cup of coffee (description of . . .), $p$ stands for a coin-description and $x = p$ is the statement that says that object $x$ satisfies the coin-description.
Of course the description is incomplete and evidently concerns an ideal machine because e.g. we have not accounted for a possible breakdown.

The same objections as against the set theoretical definition could be brought up here because it is possible to identify a function with a particular kind of set. In order to do so however we need a universe and in fact such a one that distinguishes coins e.g. according to the order in which they are presented i.e. $t(1)$ for coin 1, $t(2)$, etc. To $p(1)$ there belongs then a particular cup $c(1)$ and so on.

Then a machine is completely specified to our purpose by its behaviour, which specifies to any input function $t(n)$ the output function $c(n)$. The description $F$ of the machine should enable us to compute to any input the corresponding output. Such functions that map functions onto functions are called functionals denoted by

$$g = F(f),$$

where $f$ and $g$ are functions.

This functional description may also be applied in case of a defective machine which produces upon insertion of a coin sometimes a cup of coffee ($c$), sometimes a cup of lemon juice ($l$) and sometimes a void cup ($v$). The functional description should then enable to compute at any stage $n$ the probabilities $P_n(c)$, $P_n(l)$ and $P_n(v)$ for cups $c$, $l$ and $v$ respectively.

Evidently such a model is not quite realistic because it does not describe the output of the defective machine at work. However it gives the information on which a more realistic model may be constructed, although honestly there does not seem to be much sense in giving directions for constructing defective machines. Such models may however be useful in the analysis of defective machines in practice.

It should be noted that we need not restrict ourselves to a discrete set of stages 1, 2, 3, . . .; we could as well assume $n$ to be a continuous variable, as is well known from physical applications. In general it is advisable to postpone the decision between continuous and discrete as long as possible.

A major method of specification of functionals is by postulating relations between input and output. Because of mathematical convenience one often supposes the functional to be linear, i.e. one assumes

$$F(\alpha f + \beta g) = \alpha F(f) + \beta F(g)$$

for all functions $f$, $g$ under consideration and constants $\alpha$, $\beta$. Although such relations between input and output may contribute substantially to the identification of systems such an approach has rather strong limitations. Powerful functionals which admit a precise description fall outside the scope of functionals defined by input-output relations such as linearity. This may be illustrated by the fairly well known example of Turing machines used in the computation of functions recursive in other functions (cf. [1] or [4]).

One of the features of a Turing machine is that it has finitely many reactions to finitely many different symbols. These reactions are labeled by internal states while every reaction is accompanied by a change of state. An input of arbitrary strings of zeros and ones is converted into another such string.

Using a little mathematical technique it is possible to describe a Turing machine with great precision as a functional. It decides upon a given input whether it is acceptable and computes a result. The constructive description of the machine gives us exhaustive directions for its use. However it does not furnish a simple connection between sets of inputs and outputs such as linearity.

It may be noticed that actually many models in the practice of systems research are in the mathematical sense of the type of a functional. As an example we consider the modelcell considered by Hanken and Buijs ([2], this volume p. 11). This modelcell is an example of a functional of type

$$g = f_3 \cdot F_{f_4, f_5}(f_1, f_2),$$

where $f_1, \ldots, f_5$ and $g$ are functions of one (discrete) variable.

In this general form the $f_1$ and $f_2$ correspond to the input-functions (Hanken and Buijs prefer the term inputvariable) $Vo$ and $Po$ respectively. The function variables $f_4$ and $f_5$ correspond to the so-called decisionvariables $V$ and $Pz$, while $f_3$ corresponds to the function $MK(t)$. The functions $f_3, f_4, f_5$ act as parameters and have to be supplied separately. In fact $f_3$ (i.e. $MK$) is recursively defined in equation (1) whereas equations (2), (3), (4) may be considered to be the definition of the functional $F$. In it occur the parameters $f_4$ and $f_5$ (the decisionvariables), which apparently have to be considered as strongly connected with the system itself, while $f_3$ seems to be constant in the given context.

In this way conceived the modelcell would be typified either as $(2, 2, 1, 1)$ or if $f_3$ is not taken into account, as $(2, 2, 0, 1)$. These types have to be understood in the following way, the numbers refer in the given order to $f_1, f_2$, to $f_4, f_5$, to $f_3$ and finally to $g$.

Observe that the structure of the types differs from that given by Hanken and Buijs.

It should be noticed that in the general representation

$$g = F(f)$$

the functions $f$ and $g$ need not be restricted to the class of real functions. The domain as well as the range of both $f$ and $g$ may be various kinds of sets. For an example of a model involving rather general setfunctions cf. the mathematical treatment of information systems in [3].

Usually the conception of systems or system components as functionals does not account for a change of the systems structure with time. The requirement of change with time can be met by replacing the single functional $F$ by a one-parameter family $F_t$ of functionals (i.e. in the case of a discrete parameter a sequence of functionals).

The changes in the functional may be autonomous or induced by feed-back. We do not stress the difference because it seems that it is not correct to speak of autonomous change except inside some system, the apparently autonomous change being the result of some feed-back process on a different level, or in some other system connected with the first one.

Although feed-back which does not affect the structure of the system may be easily described mathematically either recursively as e.g. by

$$g(n+1) = F(g(n), f(n+1))$$

and in the continuous model by some integral equation, change inducing feed-back systems are more difficult to describe. Combinatorial devices for connecting simple functionals (or modelcells in the sense of Hanken and Buijs) may replace such models to a large extent but not altogether.

## Tension

In general the development of systems in time was connected with equilibrium seeking feed-back (homeostase). In recent years however the question of tension has received much attention in connection with stimulating or amplifying feed-back mechanisms. It has been observed that many systems subsist not simply by reducing tension and moving towards an equilibrium but essentially by creating tension, or at least by creating differences with the environment, resulting in changes in the system itself. When e.g. I try to solve a quadratic equation my interest is aroused if I get the wrong answer. In trying to remove the inconsistency I end up with being a different person because I have changed my behavioural potentiality in that now I know how to solve such equations in general. Tension has been removed and I look for new situations which again results in a change in the system.

Of course the picture is superficial, and one-sided. It may be applicable to separate persons but it will almost certainly fail in the question of stress in a community. Not all individuals will react in the same way and one has to introduce an individual stress parameter $\sigma(\mu)$ attached to the individual $\mu$ and for the society $S$ of individuals one has to postulate a distribution function for $\sigma$.

I have a strong impression that it will be possible to build satisfactory models for handling local as well as global situations of stress and I would strongly recommend a thorough analysis of notions of stress with an emphasis on the field of social psychology.

The first task would be to locate behaviourally input- and output actions of man in situations generally denoted as stress conditions. Then mathematical reaction-mechanisms may be set up for the individual and extended to groups.

In short I recommend an analysis of the role of tension in the activity pattern of man[3].

## References

1. Davis, M., *Computability and unsolvability*. New York (N.Y.) 1958.
2. Hanken, A. F. G. and Buijs, B. G. F., Systems analysis and business models, *Annals of Systems Research 1* (1971), 9–16.
3. Nielen, G. C. J. F., *Informatiesystemen en het besturen van ondernemingen*. Alphen a/d Rijn 1969.
4. Rogers, H. Jr., *Theory of recursive functions and effective computability*. New York (N.Y.) 1967.

[3] This subject has been chosen by a subgroup of the Systeemgroep Nederland and a progress report is planned for the next volume.

# SOME REMARKS ON AMPLIFIER MECHANISMS
## IN LIVING ORGANISMS

E. C. WASSINK[1]

## 1. Introduction

At first hand, it may seem somewhat remarkable to discuss the above subject in a group of people interested in system relationships. However, some discussions with professor Lindenmayer, and discussions at the first meeting in Utrecht (29-11-1969) had convinced me of obvious relationships between technical systems and problems and processes of plant physiology. Therefore, I have accepted the suggestion of the organizers of the present meeting to introduce a discussion on the subject mentioned in the title, although I have not been actively engaged in these problems for some time.

A first similarity between the study of living organisms and different other systems, in particular complicated technical ones, is the apparent possibility and often even the necessity to consider the organism as a black box into which the 'agent' is introduced at one side, giving some information, which inside the black box is translated in a, for the greater part unknown way into some ultimate externally observable reaction or effect.

It should be noted that in biological research, the condition of the organism at the start of the experiment almost entirely acts as a black box, i.e. as a state which is not completely known and definable. The biologist has learned to live with this circumstance which results from the complex nature of living organisms comparable to a very complex computer which is presented without a manual. He manages to get on in many cases by conditioning his experimental object in such a way that he may expect a reasonable degree of constancy of most reactive systems during the experiment. He may successfully isolate some of the reactive systems and study these by application of a suitable sequence of different intensities of environmental factors that may limit the rate of the processes concerned.

Photosynthesis of plants, for example, can be studied in this way by serial variation of light intensity, carbonic acid pressure, temperature, etc. Plant

[1] Laboratory of Plant Physiological Research, Agricultural University, Wageningen, The Netherlands. 297th Communication.

respiration can be studied similarly by conditioning the system in such a way that it reacts upon addition of organic compounds that function as respiration substrate or upon a change of oxygen pressure, etc. Naturally, these examples could be greatly multiplied. In this way one is able, albeit with great effort and very slowly, to clear a way through the black box from agent to effect, and to learn about its inner structure. It is my experience that often representants of the basic disciplines like physics and chemistry upon confrontation with the black box in living systems, have great difficulty in accepting it and often are reluctant to approach it and to cross the threshold to the biological experiment.

## 2. General properties of biological amplifier processes

As a special subject we will discuss here amplifier actions in living organisms. For a good understanding it is necessary to define the notion of amplification in biology. In general, the occurrence of a macroscopic change in a living organism (in its shape or behaviour) as a consequence of a microphysical or microchemical intervention is called amplification, in other words: the transfer of a reaction in few molecules, in the extreme case in one molecule to a great number of molecules, in the extreme case to the organism as a whole.

Some characteristic properties of amplifier processes, which appear more or less regularly, are

1. The agent 'releases' the reaction chain, which leads to the final effect. This implies that

2. Between the releasing agent and the final effect any formal and energetic discrepancy may exist.

3. As a rule, the action of the agent has finished. Between the place of action of the agent and the place or places of reaction there is a chain of processes, the so-called excitation chain or reaction chain. The objective of physiological research in this matter is the analysis of this reaction chain into its separate parts and characteristic links.

4. Generally, the effect corresponds in measure (within certain boundaries) to the quantity of the applied agent (time × intensity, time × concentration).

5. The energy necessary for the reaction is not in appreciable measure derived from the excitation (the stimulating agent), this energy playing usually only a very small role in the energy balance of the process as a whole. The reaction is established by energy which is already produced in the organism or can be produced, but is partially turned into other channels on account of the information supplied by the perceptory system.

6. Statistical phenomena (hitting chance, all-or-none law, etc.) occur at the reaction threshold in cases in which the number of initially reacting molecules per reacting unit (cell, organism) is very small.

7. A curious additional phenomenon, as yet largely unexplained, is the validity
of Weber-Fechner's law in a more or less extended range of quantities
of the active agent. This law requires the existence of a logarithmic relation
between the strength of the agent and that of the effect. In this range, the
differential sensitivity for changes in intensity of the agent is constant as a
percentage, independent of the base- or background strength of the agent.
This has been found for photobiological phenomena as well as for chemical
excitation.

Impressive examples of amplifier processes have become known from different
fields of physiology:
a, From the time course of the death curve of bacteria and other unicellular
organisms as a consequence of radiation or chemical agents. To this field
belongs also the induction of artificial mutations by radiation of chemical
agents.
b. From the action of small quantities of light on plants and animals. To this
field belong phenomena of vision, and of photostimulus physiology in plants,
e.g., the influence of small quantities of light on the course of the life cycle
and its duration in plants and animals, and other, e.g. phototropism and
phototaxis.
c. From similar actions as mentioned under b of small quantities of ergones
(usually hormones), cold-treatment etc.

In the following we will discuss some of the mentioned aspects in more detail.

## 3. Weber-Fechner's law

Already in the later decades of the last century there was great interest in the
problems of stimulus processes and discussions on Weber-Fechner's law. Most
of the characteristic properties of these processes have been formulated by the
plant physiologist Pfeffer [15] who also showed, in experimental work, that
fern spermatozoids could be persuaded to enter into capillaries containing
malic acid under the condition of a clear concentration gradient of this acid in
the surrounding liquid [14]. Pfeffer also showed that a condition for occurrence
of the reaction (the spermatozoids entering the capillaries) was a certain proportion between inner- and outer concentration of malic acid, while a threshold
concentration appeared to exist for the inner concentration if water was used
as surrounding liquid. These facts proved the validity of Weber-Fechner's law
for this biological system.

In 1888, Massart [13] proved the validity of this law for a photobiological
system, viz. the phototropic reaction of the sporangiophores of the fungus
*Phycomyces*. Massart placed a series of these sporangiophores between two

equal sources of light, and investigated the difference in illumination necessary for a specific sporangiophore in the series to bend towards one of the sources. In Massart's experiment, the required difference in intensity of light on both sides of a sporangiophore proved to be 1/5, independent of the intensity of the sources.

About 1910, discredit was thrown on these conceptions of stimulus processes by the discovery of A. H. Blaauw [2] and some others, that the quantity of energy applied determined the measure of the effect while within this quantity intensity and duration were interchangeable in a wide range. It was overlooked, however, that this result only referred to a property of the stimulus without any conclusion about the energetic and formal relation between the quantities of stimulus and effect. Pfeffer's concept was rehabilitated only after the fundamental explanation of the nature of biological amplifier processes by Otto Rahn [16] and P. Jordan [8] around 1930.

Undoubtedly, this was influenced by the development of radio engineering which had shown the possibility of suitable, quantitative amplification. Validity of Weber-Fechner's law in certain sections of the quantity of the stimulus has been found further for vision in human beings and phototactic reaction of motile photosynthesizing purple bacteria. In the range of validity of Weber-Fechner's law, i.e. where the differential sensitivity is constant, it is also maximal. At lower as well as at higher quantities of the agent, the differential sensitivity decreases. In the first case, among other things, because the probability to hit the sensitive units follows Poisson's law starts to play a role. In the second case, at high doses of the agent, because the reaction velocity is limited by factors different from the stimulating agent which then is present in excess.

The existence of these three ranges has been demonstrated by Schrammeck [18] in phototaxis of purple bacteria, and by Bouman [5] for vision of the human eye, in these two cases they extend over dose ranges of the agent (light) of very different size (figs. 1, 2). For phototaxis of purple bacteria, Manten [12] showed that the range of validity of Weber-Fechner's law almost coincides with that of light limitation of photosynthesis. If the light intensity increases so as to saturate photosynthesis, the differential sensitivity continually decreases:

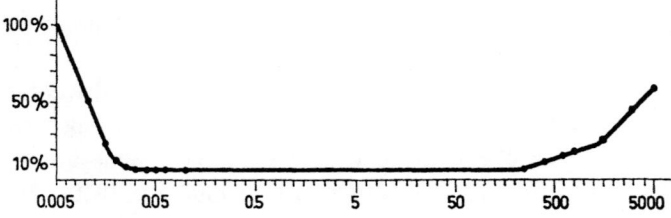

Fig. 1. Differential sensitivity for phototactic stimulation of purple bacteria in relation to background intensity (acc. to Schrammeck [18], taken from [26]).

ever greater light differences on both sides of the separation line are needed to induce the phototactic reaction on the border line. From this behaviour, Manten has drawn the plausible conclusion that probably the stimulating agent is a product of photosynthesis. This, however, does not explain the requirement

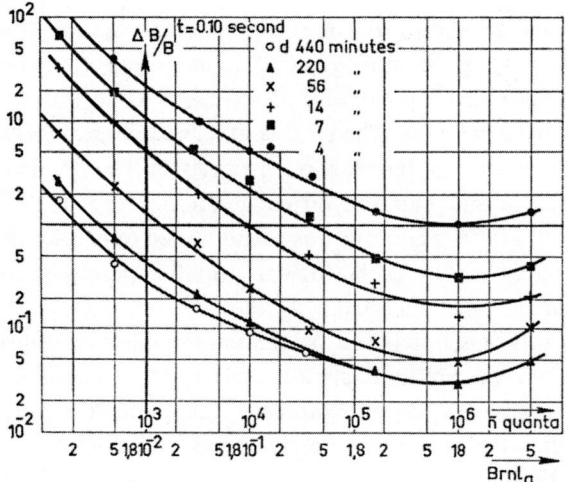

Fig. 2. Differential sensitivity of the human eye in relation to background intensity (acc. to Bouman [5], taken from [26]).

of a constant proportion between the two light intensities on both sides of the borderline in the Weber-Fechner range, because photosynthesis in this range is a linear function of light intensity, hence an equal increase is obtained at equal difference (not proportion) of two light intensities, independent of these intensities as such.

Efforts to explain this logarithmation in the reaction chain, in my opinion, have not yet led to an essential improvement of our understanding.

## 4. Time course of death curves

This is a completely different field where early observation of fundamental peculiarities has led to the assumption that in a cell very few molecules are to be found which are equipped with directory functions.

Madsen en Nyman, in 1907 [11], were among the first to observe an exponential decrease in number of survivors with time in bacteria killing experiments with various agents. The agent can be of any kind like chemical compounds, heat, injurious radiation ($\alpha$, $\beta$, $\gamma$ or ultraviolet). In 1908 and 1910, Miss Chick [6] made similar experiments and introduced statistical fluctuations in sensitivity and in probability of reaction and drew a parallel between bacteria and

molecules. She states that 'at a particular time only a proportion of the molecules (of the bacteria in case of disinfection) are temporarily in such a state as
to permit of the combination', and ascribes these phenomena to a temporary
and rhythmic change in resistance, which, in analogy to chemical processes,
may result from temporary energy changes in the constituant proteins.

It is Rahn's merit [16] to have pointed out that an appreciable spreading in
the reaction time of the reactive units only occurs if the number of initial
processes per unit is very small, e.g. 10 at most. If this number is in the order of
100, they already react almost simultaneously (fig. 3). Conversely, an experi-
mentally found spreading in reaction time may be interpreted as an indication
that only a few molecules of a specific kind play an essential role in an essential
place in the reaction chain from stimulus to effect.

In investigations of the probability of death of *coli* bacteria in $\alpha$-radiation
experiments, P. Jordan [9] estimated the size of the sensitive volume from the
active diameter of the radiation and the diameter of the $\alpha$-ionization area. In
this way he arrived at a sensitive volume, considered as a sphere, with a radius

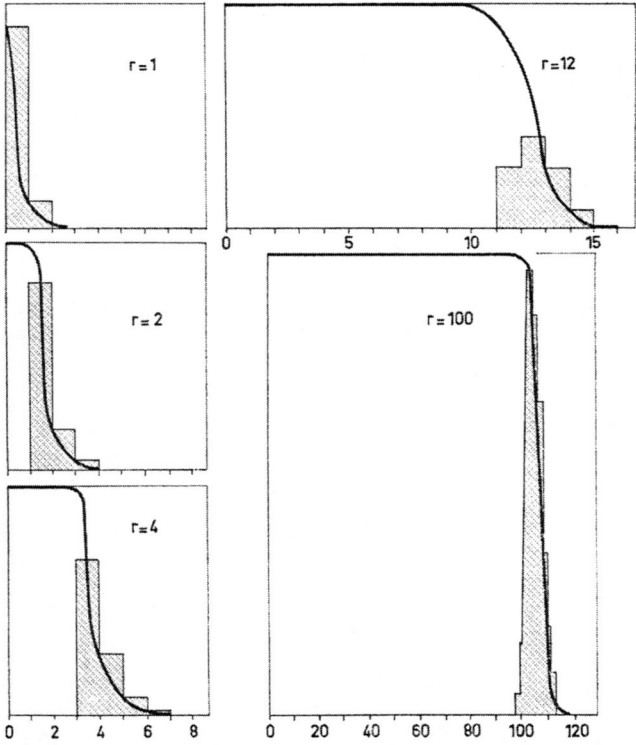

Fig. 3. Curves, showing the number of survivals (full-drawn line) and number of deaths per
time-interval (hatched) for different values of $r$ (inactivation rate 0.9); $r$ = number of
molecules required to react for death of a unit (acc. to Rahn [16], taken from [23]).

of $\sim 6.10^{-6}$ cm. If ionization occurs inside this volume, the bacterium is killed, the material content of this volume is $\sim 2.10^7$ atoms. In view of the mutual distance of the ion clusters, in general the bacterium will be killed by one ion cluster. If we put the average atomic weight of the material of the sensitive volume, which consists mainly of $H$, $C$, $N$, and $O$, at 10, the molucular weight of the sensitive volume, visualized as a single molecule, would be $\sim 10^8$. It is to be noted that Lehninger (13) gives the same order of weight for DNA-molecules. The dimensions of the bacteria are of the order of $1 \mu = 10^{-4}$ cm; so if we imagine the bacterium as a sphere, its radius would be $\sim 5.10^{-5}$, i.e. 10 times that of the sensitive volume. So the sensitive volume would be $\sim 10^{-3}$ of the total volume of the cell, and the cell would contain approximately $10^{10}$ atoms.

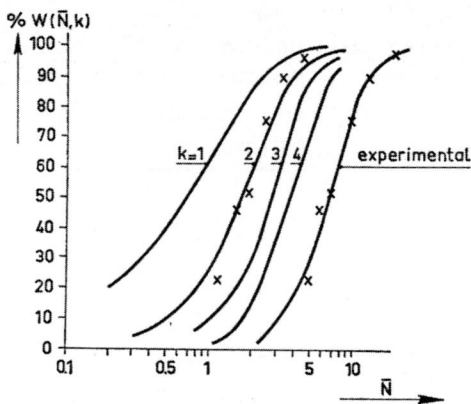

Fig. 4. Chance of observation of weak flashes of light in the human eye in relation to log energy for different values of the theoretical quantum requirement for threshold vision (k) (full-drawn lines, for different values of $k$. Crosses: experimental data fitting best the curve for $k = 2$) (acc. to Bouman [4], taken from [26]).

A field closely connected with the foregoing is the induction of artificial mutations by radiation. Here, too, the one hit rule is followed.

Killing of bacteria and the generation of mutations can also be brought about by ultraviolet light; the probability of a reaction in general, is much lower than in case of $\alpha$-particles ($\sim 10^{-3}$ times). Important in all these cases is that the reaction seems to originate from one single essential molecule.

## 5. Statistical spreading of reactions around the stimulus threshold

This phenomenon shows some relation to the previous one and is especially known from the study of photobiological processes. If the phenomenon is based on hitting one or a few molecules by a light quantum, an appreciable spreading in sensitivity occurs around the stimulus threshold, the margin between the reaction of a few and of all individuals extends over a wide energy range of the stimulus. If, on the contrary, many quanta are required (as often is so for reactions of a multicellular organ), then the range of energy doses over which the reaction from 0 to 100 percent of the individuals extends is much smaller; and the curve of percentage reacted against log energy is much steeper. All this is in accordance with Poisson's probability law.

In this way, Van der Velden [21] showed that two light quanta are required for the perception of a short (weak) light flash by the human eye. Probably these quanta are taken in by two separate rods, so that in the neighbourhood of the stimulus threshold each rod appears to be able to react upon the absorption of one light quantum (cf. Bouman [4]).

A nice feature of this experiment is the fact that its result (the shape of the reaction curve against the energy dose) is exclusively determined by the theoretical quantum requirement of each reacting unit and not by a greater or smaller individual sensitivity.

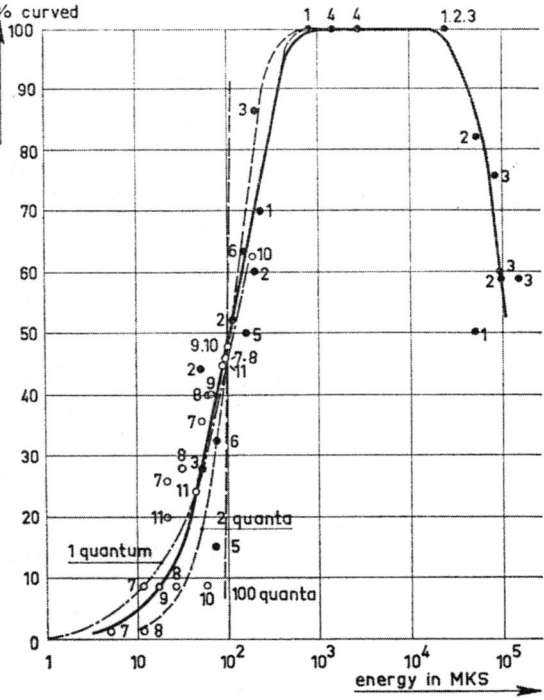

Fig. 5. Phototropic curvature of Phycomyces-sporangiophores, in relation to log energy of the stimulus (full-drawn line, according to data of Blaauw [4] indicated by various symbols). Broken lines: theoretical curves for fundamental quantum requirement per cell of 1, 2, and 100 for curvature around the stimulus threshold. Data fit best the 1-quantum curve (according to Wassink and Bouman [25], taken from [26]).

By treating in a similar way results that were long ago obtained by A. H. Blaauw [2] concerning the probability of reaction for the phototropic curvature of sporangiophores of *Phycomyces* and coleoptiles of *Avena*, Wassink and Bouman [25] concluded that the unicellular sporangiophores of *Phycomyces* can react upon one light quantum, while the multicellular *Avena*-coleoptile needs a large number (about 100); in the last case the reaction curve is much steeper (fig. 5).

## 6. Concentrations of hormonal regulators

Besides the proper nutrients which in an organism eventually form the bulk of the dry matter and occur in appreciable concentrations (e.g. proteins, carbohydrates and fats), another type of substances has been known for a long time, which occur in very low concentrations and are essential for metabolism, growth and morphogenesis of the organism. To this type, vitamines and hormones belong, often collectively denoted as ergones, regulators or 'Wirkstoffe'. Computation on the basis of effective quantities show that these compounds are effective in quantities of the order of $10^4$ to $10^6$ molecules per cell, i.e. one ergon molecule for $10^5$ to $10^9$ times its weight in protoplasmic material.

The fungus *Phycomyces* does not grow on a substrate consisting of sugar and mineral salts in contrast to many other fungi. Nor does it grow on a nutrient solution consisting of diluted and purified yeast extract. The combination of both these substrates, however, yields an excellent nutrient with very strong growth. The yeast extract apparently contains something which is indispensable for the utilization of the energy source (sugar) of the nutrient medium [22]. It has been demonstrated that the essential component is the vitamine B1 (aneurin). For the growth of 1 mg dry matter of mycelium, 5 ng aneurin proved to be necessary [18], which is a proportion of about 1 to $10^{-5}$ or $10^{-6}$. So far, the introduction of the notion of amplification is not yet necessary. The mentioned result can be explained from the principle of limiting factors: In the yeast extract the sugar supply limits growth, in the sugar solution the aneurin is limiting. There is a strong similarity with the need for inorganic nutrients on a very small scale, e.g., for trace elements (iron, manganese, copper, vanadium zinc, cobalt, etc.). The notion of amplification comes in when we realize that these small amounts of matter are required to mobilize the energy contained in the food and make it available to the metabolism of the organism. The similarity between the ergones and the inorganic micro-nutrients is that both act as active groups or co-ferments of enzymes or aid in their formation. These substances form part of the apparatus of the cell and are taken up in the pathways along which the nutrients move in the course of being processed by the cell. Co-ferments operate in metabolic pathways in general as hydrogen- or electron carriers, and form sequences of gradually increasing or decreasing redox potentials, and thus connect initial and final products of a metabolic chain. In this process, energy-rich compounds of the type of adenosine triphosphate (ATP), formed at different places in metabolism play an important role. It is clear that the concentration of the carriers may be low as compared to the quantity of metabolites to be handled (cf. e.g. [10]).

In recent years, similar chains of electron carriers have also been found in the mechanism of photosynthesis.

From this brief description of the situation, one may guess why P. Jordan [9] placed the ergones in the directory part of the living cell. Destruction of particular ergones may block large quantities of energy contained in the macro-nutrients or change their direction.

The number of ergon molecules per cell is such that destruction of one or a modest number, or slightly insufficient supply will have no observable influence on the macroscopic proceedings in the cell. The ergones are on a lower or-ganisational level of the cell than those components of which, as we have seen, the disturbance of one molecule will kill the cell or induce an all-or-none reaction.

The following example may give an impression of the proportion between numbers of light quanta and auxine-molecules in the phototropic curvature of *Avena*-coleoptiles upon a one-sided illumination of the tip. An energy of 10 ergs/$cm^2$ already leads to a strong curvature; only the extreme tip, circa 0.1 mm, is very sensitive, with the consequence that light perception takes place mainly there. Its surface may be estimated at about 0.1 $mm^2$, which, with 10 ergs/$cm^2$, receives $10^{-2}$ erg, which is about $10^9$ quanta. If one estimates a number of $10^2 - 10^3$ cells, each cell receives $10^6 - 10^7$ quanta. The energy absorbed by active pigment may be estimated to be less than 1 percent, which results in an active dose of light of about $10^4$ quanta per cell. On the other hand we know that per reacting cell about $10^4$ molecules of auxin are required for a curvature of 10°. The number of reacting cells is about 100 to 1000 times the number of perceptive ones. If one assumes that, initially, one molecule of a primary reagent is converted per quantum, the amplification factor is $10^2$ to $10^3$ for the part of the reaction chain discussed so far. Not yet included are the energies needed for the additional reactions which eventually lead to curvature of the organ (for a more extensive discussion cf. [23] ,[24], [26]).

## 7. Morphogenesis and photoperiodicity (in plants)[2]

These fields belong to a part of stimulus-physiology which has met with a large

---

[2] Even more than in the preceding sections, the references to this part are given only as easy introductions to the relevant detailed literature.

increase in knowledge in the recent decades. Small quantities of light may result in strong effects, although large energies, in the order of those required for strong photosynthesis, may have morphogenetic value and e.g. still can modify the shape of leaves (cf. [28], [29], [31], [32]).

Data about leaf growth of lettuce, collected in our laboratory lead to the assumption that the combination of an energetic component and a hormonal one decide about the outgrowth of leaves. Both depend to different degrees on duration and intensity of the light; the hormonal factor can be simulated by the addition of e.g. gibberellic acid [30]. [1].

The most important receptor system for photoperiodicity and morphogenesis in plants is phytochrome, present in almost negligible concentrations and which, upon illumination, oscillates between two forms with respective absorption maxima at 660 m$\mu$ and 730 m$\mu$. The latter is considered as the main physiologically active form (cf. e.g, [3], [27]). A brief illumination in the middle of a dark period of e.g. 14 or 16 hours may be decisive for flowering. For some plants already one treatment of this kind is effective.

Light regimes or daily exposures which lead to flowering, in annual plants at the same time stimulate the completion of the annual growth cycle and the cessation of the vegative growth. The latter continues if environmental conditions do not induce flowering, and the vegetative apparatus then becomes considerably larger than when flower-induction has taken place.

In a number of plants flowering can be induced also by certain exposures to low temperature (for further explanation cf. [26]).

Some results lead to the assumption that a flower-promoting and a flower-retarding reaction with different relations to external conditions cooperate in producing the effect.

In higher plants, the phytochromesystem is located in the leaves, where its absorption, however, is strongly dominated by that of chlorophyll (cf. [19], [20]). Experiments with regrafting of induced leaves throw some light on the processes that take place after a suitable light treatment. Zeevaart [33] showed that an induced leaf may be grafted several times upon non-induced plants, and each time again induce flowering. So, a once induced leaf seems to produce a transportable growth-stimulus for a considerable time.

## 8. Amplification and information

In general, one can say that the stimulus transfers information via the perceptive system to a more or less central directory complex in the cell, which starts the reaction chain. By specific mechanisms in the reaction chain, the number of molecules participating in the various steps is gradually or abruptly increased. In cases with a 1-hit reaction, the agent probably is of the nature of DNA.

Studies in bacterial genetics and on the composition and metabolism of *Escherichia (Bact.) coli* have produced important data on the way in which DNA directs protein synthesis via RNA and ribosomes with specific coding and transport of separate amino acids, and may change it on the basis of information received in some way e.g. by the reception of a stimulus.

Knowledge of these reaction mechanisms may be useful for a better understanding of the phenomena discussed above, because it demonstrates how a change in one or a few molecules may be amplified by a change in the pattern of biological syntheses (cf. [10]).

Finally, it may be noticed that the complication of biological structures and reaction mechanisms always appears to be greater than previously expected. This is connected with the improbability of living structures in a thermodynamical sense and with the circumstance that on the one hand they transfer a given genetic pattern and reactive mechanism fairly indefinitely to subsequent generations while on the other hand, they have to react continuously in a flexible way on subtile changes in the environment. This induces often reactions leading to deviations in the pattern of enzyme synthesis.

The impressiveness of all this becomes especially clear if we estimate the order of magnitude of information contained in living structures. After Lehninger [10], a bacterial cell of $2\mu$ diameter and $6.10^{-13}$ g weight has an information content of about $10^{12}$ bits.[3] He compares it with an extensive encyclopedia, one page of which contains $10^6$ bits, one volume of $10^3$ pages $10^9$ bits and a series of 30 volumes consequently containing about one thirtieth of the information present in one single bacterial cell. No wonder that professor L. S. Ornstein looking at mobile cells of the photoautotrophic purple sulphur bacterium *Chromatium* under the microscope, remarked that they made an impression of intelligence!

## References

1. Bensink, J., Meer, H. H. G. van der, unpublished observations.
2. Blaauw, A. H., *Rec. trav. botan. néerl.* 5, 209 (1909).
3. Borthwick, H. A., Hendricks, S. B., Toole, E. H., Toole, V. K., *Botan. Gazette* 115, 205 (1954).
4. Bouman, M. A., Thesis Univ. of Utrecht, 1949.

[3] Recently, system and information consideration have been applied to the discussion of the degree of order in ecosystems. The preliminary outcome seems to be that the information contained in the system as such is much lower than that of the composing individuals, viz. of the order of $10^5$ bits. This seems logical since the rigidity of structure and the amount of 'order' in ecosystems is definitely less than that in biological individuals. We touch here the problem that, in biological systems, the discussion of these problems requires to introduce the idea of different generalisation levels in 'system building', viz., e.g., the cell, the multicellular individual, the ecosystem, and the regional vegetation (Added in ms).

5. Bouman, M. A., *J. Opt. Soc. America*, 42, (1952) 820.
6. Chick, H., *Jl. Hygiene* 8, 92 (1908); 10, (1910) 237.
7. Hillman, W. S., Ann. Rev. Plant Physiol. 18, (1967) 301.
8. Jordan, *Naturwiss.* 20, (1932) 815.
9. Jordan, P., *Physik. Zeitschr.* 39, (1938) 345.
10. Lehninger, A. L., *Bioenergetics*, New York, Amsterdam 1965.
11. Madsen, Th., Nyman, M., *Z. Hygiene* 57, (1907) 388.
12. Manten, A., *Thesis Univ. of Utrecth*, 1948.
13. Massart, J., *Bull. Acad. royale Belgique*, 3e serie, 16, (1888) 12.
14. Pfeffer, W., *Unters. botan. Inst. Tübingen*, 1, 3, (1884) 363.
15. Pfeffer, W., *Pflanzenphysiologie*, Leipzig, I, § 3 (1897); II, § 77 (1904).
16. Rahn, O., *J. gen. Physiol.* 13, (1930) 179.
17. Schopfer, W. H., *Arch. Micribiol.* 6, (1935) 510.
18. Schrammeck, J., *Beitr. Biol. Pflanzen*, 22, (1934) 315.
19. Spruit, C. J. P., *Meded. Landbouwhogeschool Wageningen* 67, 14, 67, 15 (1967).
20. Spruit, C. J.P., *Abstracts, Europ. Photobiol. Sympos.*, Hvar, (Yugoslavia) p. VII (1967).
21. Velden, H. A. van der, *Physica* 11, (1944) 179.
22. Wassink, E. C., *Rec. trav. botan. néerl.* 31, (1934) 583.
23. Wassink, E. C., *Vakblad Biologen* 26, (1946) 13.
24. Wassink, E. C., *Rec. trav. chim. Pays-Bas* 65, (1946) 380.
25. Wassink, E. C., Bouman, M. A., *Enzymol.* 12, (1947) 193.
26. Wassink, E. C., *Proc. 1st Int. Photobiol. Congres.* Amsterdam, (1954) p. 307.
27. Wassink, E. C., Stolwijk, J. A. J., *Ann. Rev. Plant Physiol.* 7, (1956) 373.
28. Wassink, E. C., Progress in Photobiology, Christensen, B. Chr. and Buchmann, B., (eds.) *Proc. 3rd Int. Congress on Photobiology.* Copenhagen 1960, pp. 371–378.
29. Wassink, E. C., *Meded. Landbouwhogeschool Wageningen* 63, 16 (1963).
30. Wassink, E. C., *Meded. Landbouwhogeschool Wageningen* 64, 16 (1964).
31. Wassink, E. C., *Meded. Landbouwhogeschool Wageningen* 65, 15 (1965).
32. Wassink, E. C., *Meded. Landbouwhogeschool Wageningen* 69, 20 (1969).
33. Zeevaart, J. A. D., *Meded. Landbouwhogeschool Wageningen* 58, 3 (1958).